Holt McDougal Mathematics

Course 2
Chapter 7 Resource Book

HOLT McDOUGAL
a division of Houghton Mifflin Harcourt

Printed in the United States of America

ISBN 13: 978-0-55-400740-3
ISBN 10: 0-55-400740-1

2 3 4 5 170 12 11 10 09

Contents

Description of Contents

Family Involvement Pages

The Chapter Resource Book includes a set of family involvement pages for each section. The family involvement pages consist of the following items:

- **Family Letter**, a two-page letter describing the math that the student will study in each section. A list of vocabulary words from the lessons is included, and some worked-out examples are provided.
- **At-Home Practice**, a one-page worksheet with problems drawn from the content described in the Family Letter. Answers are provided on the page so that the student and his or her family can check this work at home.
- **Family Fun**, a one-page activity sheet that the student and his or her family can work on together.

Practice A, B, and C

There are three practice worksheets for every lesson. All of these reinforce the content of the lesson. Practice B is shown in the Teacher's Edition and is appropriate for the on-level student. It is also available as a workbook (Homework & Practice Workbook)

Practice A is easier than Practice B but still practices the content of the lesson. Practice C is more challenging than Practice B.

Review for Mastery

The Review for Mastery worksheet (one per lesson) provides an alternate way to teach or review the main concepts of the lesson. This worksheet is one or two pages long and is shown in the Teacher's Edition.

Challenge

The Challenge worksheet (one per lesson) enhances critical thinking skills and extends the lesson. This worksheet is shown in the Teacher's Edition.

Problem Solving

The Problem Solving worksheet (one per lesson) provides practice in problem solving and opportunities for real-world applications and for interdisciplinary connections. There are both multiple choice and short response problems. This worksheet is shown in the Teacher's Edition.

Reading Strategies

The Reading Strategies worksheet (one per lesson) provides tools to help the student master math vocabulary or symbols.

Puzzles, Twisters & Teasers

The Puzzles, Twisters & Teasers worksheet (one per lesson) provides fun practice while reinforcing the content of the lesson.

CHAPTER 7

Family Letter

7A Organizing and Displaying Data

Dear Family,

As the student learns about organizing data, he or she will use both a **frequency table** and a **stem-and-leaf plot.** These data displays will help the student organize large sets of data.

Make a cumulative frequency table of the data.

The table shows the number of bikes sold at a toy store within a one-year time period.

Month	Sold	Month	Sold
Jan	12	July	54
Feb	19	Aug	60
Mar	25	Sep	35
Apr	28	Oct	32
May	38	Nov	38
June	55	Dec	75

Step 1 Choose equal intervals.

Step 2 Find the number of data values in each interval. Write these in the "Frequency" column.

Sales Range	Frequency	Cumulative Frequency
0–19	2	2
20–39	6	8
40–59	2	10
60–79	2	12

Step 3 Cumulative frequency is found by adding all the frequency values above or in that row.

The student will learn how to use measures of central tendency to describe trends in a set of data. An example is shown below.

Find the mean, median, mode, and range of the set.

26, 27, 28, 22, 28, 26, 26, 22, 20

Mean

26 + 27 + 28 + 22 + 28 + 26 + 26 + 22 + 20 = 225 — Add the values.

225 ÷ 9 = 25 — Divide the sum by the number of items.

The mean is 25.

Median

Arrange the values in order.

20, 22, 22, 26, 26, 26, 27, 28, 28 — Choose the number in the middle of the data set.

The median is 26.

Mode

The mode is 26. — The value 26 occurs 3 times.

Range

28 − 20 = 8 — Subtract the least value from the greatest value.

The range is 8.

Vocabulary

These are the math words we are learning:

bar graph a display of data in which horizontal or vertical bars are used to compare quantities

box-and-whisker plot a method of showing how data is distributed by using the median, the upper and lower quartiles, and the greatest and least values

circle graph sometimes called a pie chart, this graph shows how a set of data is divided into parts

cumulative frequency a running total of the number of data items in each column of a frequency table

double-bar graph a bar graph that compares two sets of related data

frequency table a way to organize data values into categories or groups

histogram a bar graph that shows the frequency of data within equal intervals

interquartile range the difference between the upper and lower quartiles

line plot a way to display data using a number line to show frequency of values

Family Letter

7A Organizing and Displaying Data continued

The student will be interpreting and constructing many different types of graphs used to display data. A **histogram** shows the frequency of data within equal intervals.

The table shows the number of hours per week that students spend exercising. Make a histogram.

<table>
<tr><th colspan="10">Number of Hours Spent Exercising</th></tr>
<tr><td>Hours</td><td>1</td><td>2</td><td>3</td><td>4</td><td>5</td><td>6</td><td>7</td><td>8</td><td>9</td></tr>
<tr><td>Number of Students</td><td>||</td><td>𝍸 |</td><td>||
||</td><td>𝍸 |</td><td>||</td><td>|||</td><td>||</td><td>||||</td><td>|</td></tr>
</table>

Step 1 Make a frequency table using equal intervals.

Step 2 Choose an appropriate scale and interval for the vertical axis. The greatest value on the scale should be at least as great as the greatest frequency.

Step 3 Draw a bar for each interval. The height of the bar is the frequency for that interval. Bars must touch but not overlap.

Step 4 Label both axes. Give the graph a title.

Number of Hours	Frequency
1–3	12
4–6	11
7–9	7

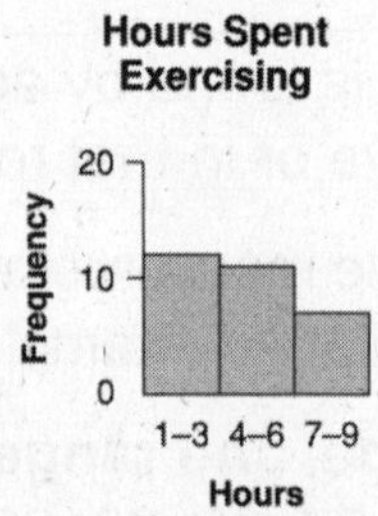

Another way to display data is a **box-and-whisker plot.** The student will determine how data is distributed by finding the median, the least and greatest values, and the upper and lower quartiles of the data set.

Shown below is a box-and-whisker plot with a median of 25, a least value of 15, a greatest value of 32, a lower quartile of 20.5, and an upper quartile of 30.

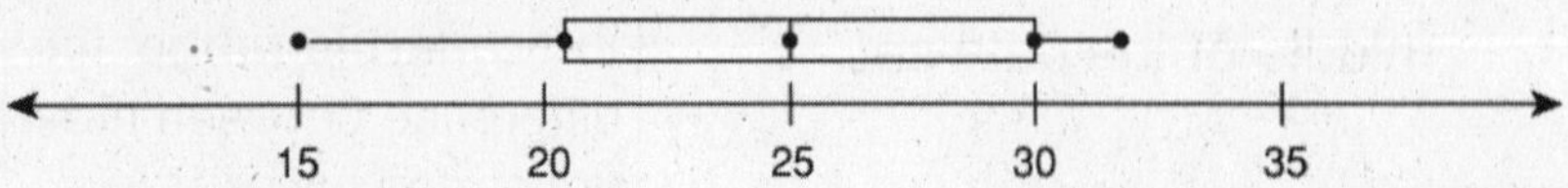

As the student continues to study data and the ways data can be displayed, encourage him or her to be aware of the ways data is presented in the media.

Sincerely,

lower quartile the median of the lower half of a data set

mean the sum of data values divided by the number of data items

median the middle value of an odd number of items arranged in order, or for an even number of items, the mean of the two middle values

mode the value or values that occur most often

outlier an extreme value in a data set

range the difference between the greatest and least values

sector a slice of a circle graph that represents one part of the entire data set

stem-and-leaf plot a way to show how often data values occur and how they are distributed; each leaf on the plot represents the right-hand digit and each stem represents the left-hand digits in a data value

upper quartile the median of the upper half of a data set

Name ______________________ Date ______________ Class ______________

CHAPTER 7

At-Home Practice

7A Organizing and Displaying Data

Find the mean, median, mode, and range of the data set.

1.

Game Date	Oct 2	Oct 6	Oct 13	Oct 18	Oct 25
Goals Scored	3	6	4	3	2

____

____

Use the table to fill in the missing values for the frequency table. The table shows the number of minutes each cyclist rode.

Cyclist	Minutes	Cyclist	Minutes
Jim	13	Ron	55
Al	21	Ray	60
Wendy	60	Tim	22
Angel	14	Mark	44
Bill	35	Shirley	16
Luis	29	Dan	61

	Minutes Rode	Frequency	Cumulative Frequency
	0–19	3	3
2.	20–39		7
3.	40–59	2	
	60–79	3	12

The bar graph shows the sales of two stores. Use the graph for Exercises 4–5.

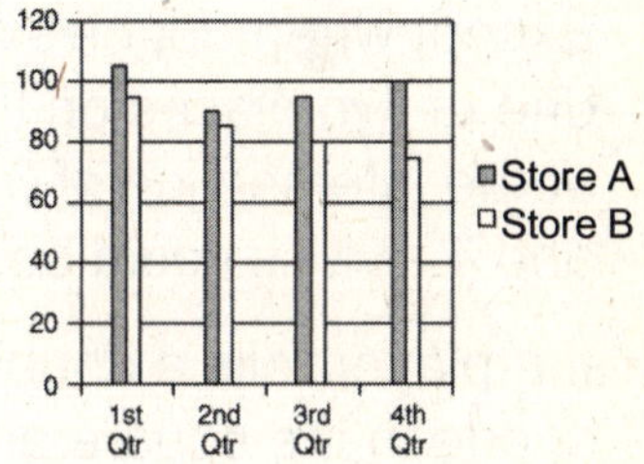

4. In which quarter did Store B have the least amount in sales? ______________________

5. Who had the most sales in the first quarter?

____

The pie chart shows the cases of rabies in the United States. Use the graph for Exercises 6–7.

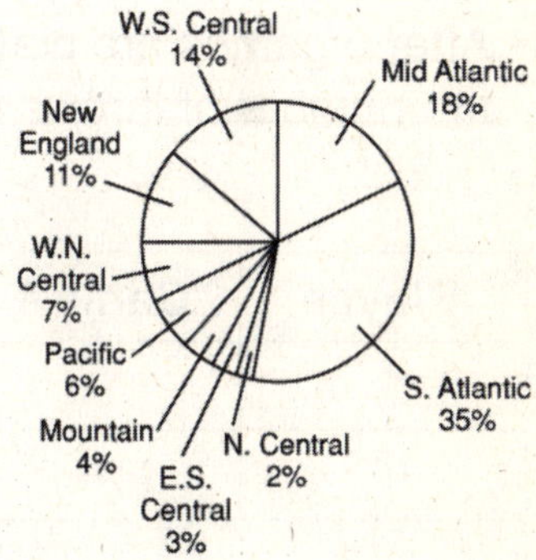

6. Which region had the most reports of rabies?

____

7. Which two regions make up more than 50% of the reported cases?

____

The table shows the students who earned straight A's in each term. Use the box-and-whisker plot and table for Exercises 8–9.

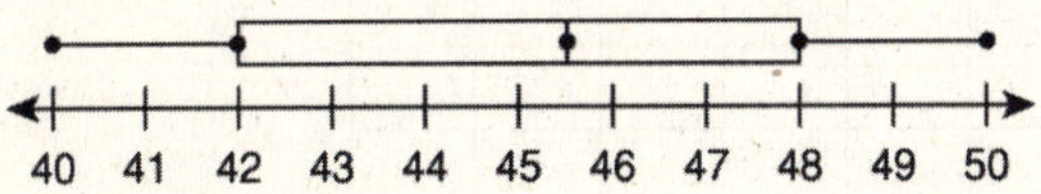

Term	1	2	3	4	5	6
Number of students	50	45	46	40	42	48

8. What is the median?

____

9. What is the range of the data?

____

Answers: **1.** Range 4, Mean 3.6, Median 3, Mode 3, **2.** 4 **3.** 9 **4.** 4th Quarter **5.** Store A **6.** S. Atlantic **7.** S. Atlantic and Mid-Atlantic **8.** 45.5 **9.** 10

Name ______________________ Date ______________ Class ______________

CHAPTER 7

Family Fun

Graphing Good Times

Materials

2 number cubes
Math fact sheets
Pencil
Graph paper

Objective

Be the first to complete a row or a column in a math fact sheet. Players keep a tally of the number of times cubes are tossed in the game.

Directions

- Each player makes a math fact sheet at the beginning of each new game. (See below for an example of the math fact sheet.)
- One player tosses the number cubes. Keep a tally of the number of times the cubes are tossed.
- The two numbers on the cubes are factors. The player enters and shades the product in the correct square on the math fact sheet. Except when doubles are rolled, each product may be entered in one of two places on the sheet. (If a 6 and 4 are tossed, the square at the intersection of 6 across and 4 down or the square at 6 down and 4 across could be shaded.)
- If a player rolls a double and knows the product, that player can shade in the square and roll again.
- The first player to complete a row or a column is the winner.
- After 5 games are played, players can make a bar graph showing the number of throws per game.

Number of Throws Tally

Player	Game 1	Game 2	Game 3	Game 4	Game 5

Math Fact Sheets

Player 1 ←

	1	2	3	4	5	6
1						
2						
3						
4						
5						
6						

Player 2 →

	1	2	3	4	5	6
1						
2						
3						
4						
5						
6						

Name ______________________ Date ______________ Class ______________

LESSON 7-1

Practice A

Frequency Tables, Stem-and-Leaf Plots, and Line Plots

The table shows normal monthly temperatures for Tampa, Florida, for each month of the year.

Normal Monthly Temperatures in Tampa						
Month	**January**	**February**	**March**	**April**	**May**	**June**
Temp. (°F)	61	63	67	72	78	82

Month	**July**	**August**	**September**	**October**	**November**	**December**
Temp. (°F)	83	83	82	76	69	63

1. Complete the cumulative frequency table for the data.

Normal Monthly Temperatures in Tampa		
Temperature (°F)	**Frequency**	**Cumulative Frequency**
60–69		
70–79		
80–89		

2. How many months had a normal temperature of less than 80°? ______________

3. Complete a stem-and-leaf plot of the data.

Normal Monthly Temperatures in Tampa

Stems	**Leaves**
6	
7	
8	

Key:

4. How many months had a normal temperature in the 60s? ______________

5. Complete a line plot of the data.

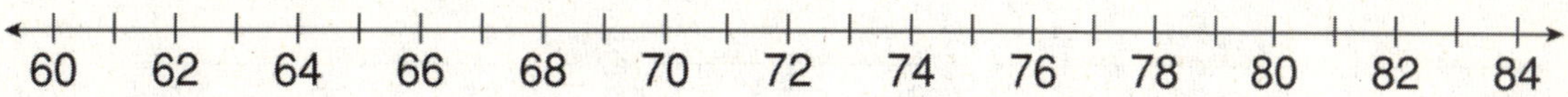

Name ____________________ Date ______________ Class ______________

LESSON 7-1

Practice B

Frequency Tables, Stem-and-Leaf Plots, and Line Plots

The table shows the heights of students in Ms. Blaire's class. Use the table for Exercises 1 and 2.

	Height (in.)
Males	60, 45, 48, 57, 62, 59, 57, 60, 56, 58, 61, 52, 55
Females	49, 52, 56, 48, 51, 60, 47, 53, 55, 58, 54

1. Make a cumulative frequency table of the data.

Heights of Students

Height (in.)	**Frequency**	**Cumulative Frequency**

2. How many of the students were less than 60 in tall? ______________

3. Make a stem-and-leaf plot of the data.

Heights of Students

Stem	**Leaves**

Key:

4. How many of the students were less than 50 inches tall? ______________

5. Make a line plot of the data.

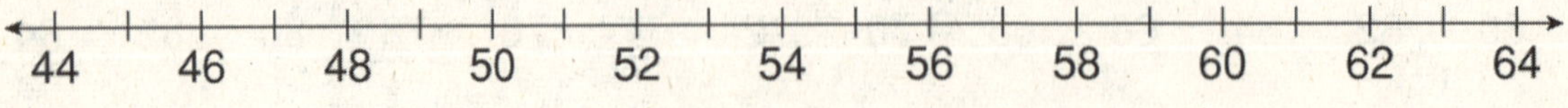

6. Which height occurred the greatest number of times? ______________

Name ______________________ Date ______________ Class ______________

LESSON 7-1

Practice C

Frequency Tables, Stem-and-Leaf Plots, and Line Plots

The list below shows the ages of all the presidents inaugurated in the twentieth century.

42, 51, 56, 55, 51, 54, 51, 60, 62, 43, 55, 56, 61, 52, 69, 64, 46

1. Make a cumulative frequency table of the data.

2. Make a stem-and-leaf plot of the data.

Key:

3. Make a line plot of the data.

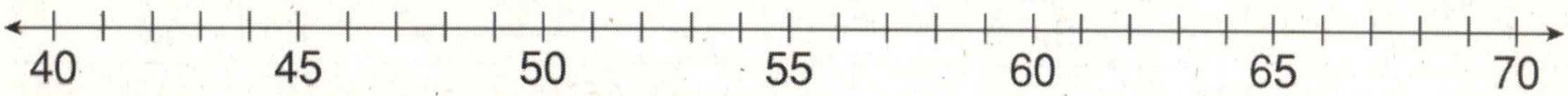

4. How many of these presidents were in their 50s when they took office? ________

5. Which of the following is the most likely source of the data in the stem-and-leaf plot? Explain your answer.

   a. class sizes in a middle school

   b. speeds in miles per hour of cars on a highway

   c. number of hours worked in a week by part-time employees of Toy Town

Stems	Leaves
5	1 1 2 2 3 4 5 6 8 8
6	0 0 0 0 4 1 8
7	1 2

Key: 5 | 8 means 58

____

____

Name ______________________ Date ______________ Class ______________

LESSON 7-1

Review for Mastery

Frequency Tables, Stem-and-Leaf Plots, and Line Plots

A **frequency table** organizes data into groups.

A **cumulative frequency** table includes a running total of the frequencies of all the previous groups.

The list shows the pulse rate of 12 students.

65, 63, 72, 88, 90, 82, 80, 75, 68, 72, 74, 84

You can use the data to make a cumulative frequency table. Here's how.

Step 1: Choose an interval. Each interval must be the same size. Often, 5 or 10 is a good interval length. Write the intervals in the first column.

Step 2: Find the number of values in each interval. Write the number in the middle column.

Step 3: Add the values to find the cumulative frequency. Write the number in the last column.

intervals of 10

Pulse Rates

Pulse Rate	Frequency	Cumulative Frequency
60–69	3	3
70–79	4	4 + 3 = 7
80–89	4	4 + 7 = 11
90–99	1	1 + 11 = 12

the number of students with pulse rates between 80 and 89

the sum of the frequency column is equal to the last number in the cumulative frequency column.

1. The list below shows the highest temperatures (°F) ever recorded in all the states that begin with the letters M or N in the United States. Use the data to complete the cumulative frequency table.

   105°, 109°, 107°, 112°, 114°, 115°, 118°, 117°, 118°, 125°, 106°, 110°, 122°, 108°, 110°, 121°

Highest Temperatures in M and N States

Temperature (°F)	Frequency	Cumulative Frequency
105–109	5	5
125–129		

Name ______________________ Date ______________ Class ______________

LESSON 7-1

Review for Mastery

Frequency Tables, Stem-and-Leaf Plots, and Line Plots (continued)

A **stem-and-leaf plot** shows how often data values occur and how they are distributed. Numbers are separated into two parts: the stem and the leaves. The stem shows the leftmost digit in each number, and the leaves show the rightmost digit in each number.

You can use the pulse rate data to make a stem-and-leaf plot. Here's how.

65, 63, 72, 88, 90, 82, 80, 75, 68, 72, 74, 84

Step 1: List the stems in the left column of the plot.

Step 2: List the leaves in order in each row that corresponds to the stem of the number.

Step 3: Write a key and a title.

The stems are the left digits of each number.

Pulse Rates

Stems	Leaves
6	3 5 8
7	2 2 4 5
8	0 2 4 8
9	0

The leaves are the right digits of each number.

Key: 7 | 2 means 72

The first stem (6) has three leaves (3, 5, and 8). This represents the data values 63, 65, and 68.

2. The list below shows the number of stories of some of the tallest buildings in Houston, Texas. Use the data to complete the stem-and-leaf plot.

   55, 52, 53, 64, 56, 71, 50, 46, 49, 75
   50, 49, 46, 36, 47, 45, 47, 36, 44, 42

Tallest Buildings in Houston

Stems	Leaves
3	6
4	

Key:

Name ____________________ Date ____________ Class ____________

LESSON 7-1

Challenge

Read From the Middle Out

In a double stem-and-leaf plot, the stem is in the middle and the leaves are on both sides. You read from the middle to the left for the left data and the middle to the right for the right data.

The double stem-and-leaf plot compares the average monthly temperatures in Bloomington, Indiana, and Richmond, Virginia, in degrees Fahrenheit.

Average Monthly Temperatures (°F)

Bloomington Leaves	Stems	**Richmond** Leaves
7	2	
3 1	3	8
5 2	4	0 0 8 9
6 3	5	7 9
7 3	6	6
6 4 2	7	0 4 7 8

Key: 7 | 2 | means 27 °F

Key: | 3 | 8 means 38 °F

Use the double stem-and-leaf plot above to answer the questions.

1. What is the greatest average monthly temperature in Bloomington? in Richmond?

____

2. Which city had the lowest monthly temperature?

____

3. Which city has more months with monthly temperatures below 30 °F?

____

____

4. Which city has more variation in temperatures? Explain.

____

____

5. What average temperature did Richmond have twice?

____

6. How many months was the average monthly temperature above 68° in Richmond?

____

7. What was the lowest average monthly temperature in Bloomington?

____

8. Compare the highest average monthly temperature in Bloomington to the highest average monthly temperature in Richmond.

____

____

____

Name ______________________ Date ______________ Class ______________

LESSON 7-1

Problem Solving

Frequency Tables, Stem-and-Leaf Plots, and Line Plots

Write the correct answer.

The table shows the time in minutes that Naima talked on the phone during the last 3 weeks.

Phone Time (min)

	Mon	Tues	Wed	Thurs	Fri	Sat	Sun
Week 1	12	15	25	45	52	30	31
Week 2	22	25	46	51	10	19	33
Week 3	44	21	30	20	10	24	52

1. Naima made a cumulative frequency table of the data using equal intervals. What number would she write in the frequency column for the interval 11–20 minutes?

____

2. Naima made a line plot of the data. Which numbers had more than one X above them?

____

3. If Naima makes a stem-and-leaf plot, which stem has the most leaves? What are they?

____

____

4. In the stem-and-leaf plot, which stems have the same number of leaves?

____

____

The list shows Hank Aaron's season home run totals. Make a cumulative frequency table, stem-and-leaf plot, and a line plot for the data. Then use the data to solve problems 5–8.

13, 27, 26, 44, 30, 39, 40, 34, 45, 44, 24, 32
44, 39, 29, 44, 38, 47, 34, 40, 20, 12, 10

5. In a cumulative frequency table of the data, what number belongs in the frequency column for interval 40-44?

   A 5   C 8
   B 6   D 14

6. In a cumulative frequency table of the data, what number belongs in the frequency column for interval 25-29?

   F 3   H 6
   G 5   J 8

7. In a stem-and-leaf plot of the data, how many stems do you need?

   A 1   C 3
   B 2   D 4

8. In a line plot of the data, which number would have 4 x's above it?

   F 34   H 44
   G 40   J 45

Name ______________________ Date ______________ Class ______________

LESSON 7-1

Reading Strategies

Use Graphic Aids

To the right is a listing of points scored in each of the league's basketball games.

Creating a **frequency table** is one way to organize the scores. A frequency table organizes scores by how often they occur.

28 48 34 50 47 35 40 37 36
55 43 39 43 34 52

League Basketball Scores

Points Scored	Frequency
20–29	1
30–39	6
40–49	5
50–59	3

Answer these questions about the frequency table.

1. How are the scores organized?

____

The **frequency,** or how often the scores occur, is recorded in the second column.

2. In how many games were 50 or more points scored?

____

3. In how many games were between 30 and 39 points scored? __________

A **stem-and-leaf plot** is another way to organize the data. Since the scores are two-digit numbers, the **stem** represents the tens digits, and the **leaves** represent the ones in each score.

League Basketball Scores

Stems (tens)	**Leaves** (ones)
2	8
3	4 4 5 6 7 9
4	0 3 3 7 8
5	0 2 5

Key: 2 | 8 means 28

Stem → 3

4 4 5 6 7 9 ← **Leaves**

Data in this row: 50, 52, 55 → 5

Answer each question about the stem-and-leaf plot.

4. Are the stems on the right or the left in a stem-and-leaf plot? __________________

5. How many stems are in this plot? ______________________________

6. What number does 3|6 represent? ______________________________

Name ______________________ Date ______________ Class ______________

LESSON 7-1

Puzzles, Twisters & Teasers

The Plot Thickens!

Have you ever wondered about traveling trees? Finish this page to discover the answers to two important traveling tree questions! The list shows the number of minutes people spend watching television each day. Complete the cumulative frequency table of the data. Write the corresponding letters above the correct answers to solve the riddle.

10, 10, 15, 18, 25, 28, 32, 35, 35, 45, 50, 55, 57, 59

Minutes	Frequency	Cumulative Frequency
10–20		A
21–30		E
31–40		L
41–50		S
51–60		V

What does a tree do when it wants to go away?

It ____ ____ ____ ____ ____ ____ .
9 6 4 14 6 11

Use the list to make a stem-and-leaf plot. Write each letter on the line above the correct answer at the bottom of the page to solve the riddle.

Stems	Leaves
	K
	N
	R
	T
	U

Where does a traveling tree pack its things?

In its ____ ____ ____ ____ ____ .
45 32 57 28 15

Name ________________ Date ____________ Class ____________

LESSON 7-2

Practice A

Mean, Median, Mode, and Range

Find the mean, median, mode, and range.

1. 5, 8, 5, 9, 3

____

2. 58, 54, 60, 56, 52

____

3. 18, 17, 21, 18, 26

____

4. 60, 20, 40, 10, 50, 30

____

The line plot below shows the number of kilometers Clara ran each day for 14 days. Use the line plot for Exercises 5 and 6.

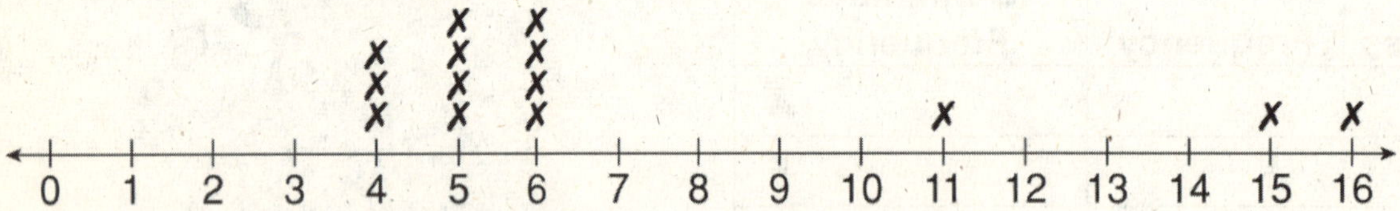

5. Find the mean, median and mode for the set of data.

____

6. Which measure of central tendency best describes the data? Explain your answer.

____

____

Use the data set to answer the questions.

22, 18, 15, 19, 61, 21

7. What is the outlier? ___________

8. How does the outlier affect the mean, median, and mode?

____

____

9. Which measure of central tendency best describes the data set with the outlier?

____

10. Which measure of central tendency best describes the data set without the outlier?

____

Name ______________________ Date ______________ Class ______________

LESSON 7-2

Practice B

Mean, Median, Mode, and Range

Find the mean, median, mode, and range of each data set.

1. 46, 35, 23, 37, 29, 53, 43

2. 72, 56, 47, 69, 75, 48, 56, 57

3. 19, 11, 80, 19, 27, 19, 10, 25, 15

4. 7, 8, 20, 6, 9, 11, 10, 8, 9, 8

5. The line plot shows the number of hours 15 students said they spent on homework in one week. Which measure of central tendency best describes the data? Justify your answer.

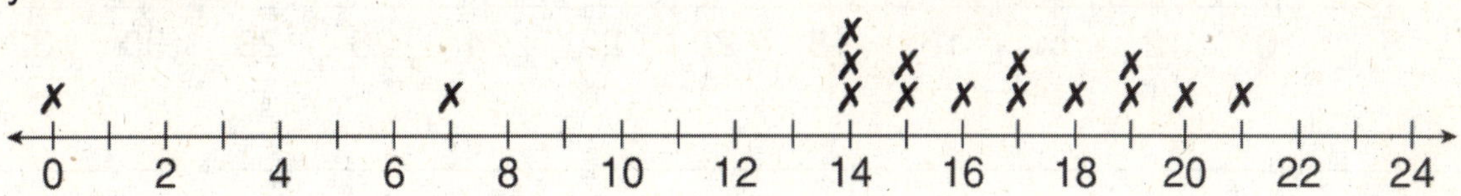

Identify the outlier in each data set. Then determine how the outlier affects the mean, median, and mode of the data. Then tell which measure of central tendency best describes the data with and without the outlier.

6. 14, 16, 13, 15, 5, 16, 12

7. 48, 46, 52, 92, 57, 58, 52, 61, 56

Name ________________ Date ________ Class ________

LESSON 7-2

Practice C

Mean, Median, Mode, and Range

Find the mean, median, mode, and range of each data set.

1. 21, 14, 17, 16, 23, 17, 15, 13

2. 44, 56, 38, 55, 62, 56, 48, 62, 56

3. The line plot shows Omar's point totals for 20 basketball games. Which measure of central tendency best describes the data? Justify your answer.

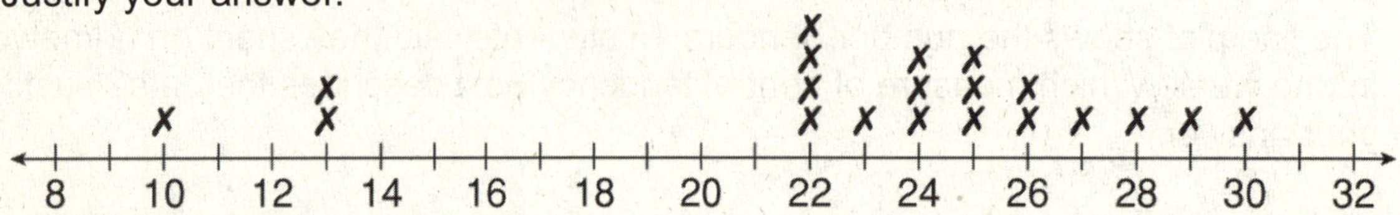

Identify the outlier in each data set. Then determine how the outlier affects the mean, median, and mode of the data. Then tell which measure of central tendency best describes the data with and without the outlier.

4. 26, 33, 36, 26, 30, 26, 10, 33

5. 59, 56, 62, 103, 61, 67, 62, 52, 63

6. Bart has these quiz scores: 78, 76, 74, 80, 82, 78, 82, 75, 77, 82. Bart tells his parents that his typical quiz score is an 82. Which measure of central tendency did Bart use to describe his typical score? Do you think Bart's measure of central tendency best described his typical score? Explain.

Name ______________________ Date ______________ Class ______________

LESSON 7-2

Review for Mastery

Mean, Median, Mode, and Range

Measures of central tendency show what the middle of a data set looks like. The measures of central tendency are the *mean, median,* and *mode.*

Find the mean, median, mode, and range of 8, 3, 5, 4, 1, and 3.

Find the **mean**.
The mean is the sum of the values divided by the number of values in the data set.
1 + 3 + 3 + 4 + 5 + 8 = 24
24 ÷ 6 = 4
mean = 4

Find the **range**.
Find the difference between the least and greatest values.
8 − 1 = 7
range = 7

List in order: 1, 3, 3, 4, 5, 8

Find the **mode**.
The mode is the value that occurs most often.
Sometimes there is no mode.
mode = 3

Find the **median**.
The median is the middle value.
median = 3.5

Find the range, mean, median, and mode of each data set.

1. 6, 5, 3, 6, 8

____

2. 12, 15, 17, 9, 17

____

3. 26, 35, 23, 27, 19, 23

____

4. 7, 6, 13, 16, 15, 9

____

5. 42, 38, 45, 42, 43

____

6. 51, 62, 68, 55, 68, 62

____

7. **Monthly Low Temperatures**

Month	**Jun.**	**Jul.**	**Aug.**	**Sept.**	**Oct.**
Temperature (°F)	44	41	47	42	36

____

Name ______________________ Date ______________ Class ______________

LESSON 7-2

Review for Mastery

Mean, Median, Mode, and Range (continued)

An **outlier** is a value that is much greater than or much less than the other values in a data set.

How does the outlier affect the mean, median, and mode of the data?

7, 9, 30, 9, 5, 6

Write the data in order and identify the outlier.

5, 6, 7, 9, 9, **30** ← outlier

With the Outlier	Without the Outlier
Find the mean. 7 + 9 + 30 + 9 + 5 + 6 = 66 66 ÷ 6 = 11 The mean is 11.	Find the mean. 7 + 9 + 9 + 5 + 6 = 36 36 ÷ 5 = 7.2 The mean is 7.2.
Find the median. 5, 6, 7, 9, 9, 30 7 + 9 = 16 16 ÷ 2 = 8 The median is 8.	Find the median. 5, 6, 7, 9, 9 The median is 7.
The mode is 9.	The mode is 9.

8. How does the outlier affect the mean? ______________________

9. How does the outlier affect the median? ______________________

10. How does the outlier affect the mode? ______________________

To choose the measure of central tendency that best describes a set of data:

- Choose the measure that is closest to the greatest number of values in the data set, OR
- If there is an outlier, think about how it affects the mean and the median. Choose the measure that is affected least by the outlier.

Use the data set to answer the questions.

4, 6, 3, 6, 25, 3, 2

11. Is there an outlier? If so, what is it? ______________

12. How does the outlier affect the mean and the median?

____

13. Which measure of central tendency best describes the data? Explain your answer.

____

____

Name ______________________ Date ______________ Class ______________

LESSON 7-2

Challenge

These Puzzles Are Mean

Solve each puzzle.

1. There are 6 whole numbers in a set of numbers. The least number is 8, and the greatest number is 14. The mean, the median, and the mode are 11. What are the numbers?

____

2. There are 7 whole numbers in a set of numbers. The least number is 10, and the greatest number is 20. The median is 16, and the mode is 12. The mean is 15. What are the numbers?

____

3. There are 8 whole numbers in a set of numbers. The greatest number is 17, and the range is 9. The median and the mean are 12, but 12 is not in the data set. The modes are 9 and 14. What are the numbers?

____

4. The mean of a data set of 6 numbers is 8. The mean of a different data set of 6 numbers is 20. What is the mean of the combined data sets? ____________

5. Find the mean of 7 numbers if the mean of the first 4 numbers is 5 and the mean of the last 3 numbers is 12. What is the mean of the combined data sets? ____________

6. The mean of a data set of 3 numbers is 12. The mean of a data set of 9 numbers is 40. What is the mean of the combined data sets? ____________

7. Sasha needs an average of 30 points to move to the next level in her competition. Her scores in the first three events are 28, 35, and 30. What is the lowest score she can score in her last event to move to the next level of competition? ____________

8. Lars has a score of 89 for each of his first 3 science quizzes. The score on his fourth quiz is 92. What score does he need on his fifth quiz to have an average of 90? ____________

9. Jake has a 95 average in math after 4 quizzes. Then he got 0 on the next quiz after being absent. There are 2 more quizzes. What average grade does he need on these last 2 quizzes to keep his average at least 85?

____

Name ______________________ Date ______________ Class ______________

LESSON 7-2

Problem Solving

Mean, Median, Mode, and Range

Write the correct answer.

The table to the right shows the leading shot blockers in the WNBA during the 2003 season.

Player	Shots Blocked
Margo Dydek	100
Lauren Jackson	64
Lisa Leslie	63
Ruth Riley	58
Michelle Snow	62

1. What is the range of this set of data?

____

2. What are the mean, median, and mode of this set of data?

____

3. What is the outlier in this set of data?

____

4. How does the outlier affect the mean and the median?

____

____

____

5. Which measure of central tendency best describes the set of data with the outlier? Explain.

____

____

____

Choose the letter for the best answer.

In a 100-meter dash, the first 5 racers finished with the following times: 11.6 seconds, 13.4 seconds, 10.8 seconds, 11.8 seconds, and 13.4 seconds.

6. Which measure of central tendency for this set of data is 12.2 seconds?
   - A mean
   - B median
   - C mode
   - D none of the above

7. Which measure of central tendency for this set of data is 11.8 seconds?
   - F mean
   - G median
   - H mode
   - J none of the above

8. What is the mode for this set of data?
   - A 10.8 seconds
   - B 11.8 seconds
   - C 13.4 seconds
   - D none of the above

9. The sixth racer finished with a time of 16.4 seconds. How will that affect the mean for this set of data?
   - F decrease it by 0.7 second
   - G increase it by 0.7 second
   - H increase it by 3.28 seconds
   - J does not affect the mean

Name ______________________ Date ______________ Class ______________

LESSON 7-2

Reading Strategies

Understanding Vocabulary

There are different ways to look at data. The **mean**, the **median**, the **mode**, and the **range** help us understand the information we gather.

The mean, or **average,** is the sum of the values divided by the total number of values. Example: Ages in years ⟶ 18 11 6 12 18

The average age is (18 + 11 + 6 + 12 + 18) ÷ 5 = 13 years

The **median** is the middle value in a set of data, when the values are listed in order. The median is 12.

6 11 **12** 18 18

The **mode** is the most frequent value in a set of data. The mode is 18.

6 11 12 **18** **18**

The **range** indicates the spread of the data. It is the difference between the least and greatest values in a set of data.

18 – 6 = **12**. The range is 12.

These are Jack's test scores for 5 math tests: 83, 79, 92, 95, 83. Use this data to help Jack look at his scores. Write "mean," "median," "mode" or "range" to answer the following questions.

1. 83 occurs more than any of the other values. Which measure is 83?

   ____

2. Jack added his test scores and divided by 5 to find which measure?

   ____

3. Jack found that the difference between his highest and lowest score was 16 points. What is this measure called?

   ____

4. Jack found that the middle value is 83. What is that value called?

   ____

Name ________________ Date ________________ Class ________________

LESSON 7-2

Puzzles, Twisters & Teasers

Stuck in the Middle!

Ever wonder what number in a data set sticks out like a sore thumb? To find out, use the data sets below to answer each question. Fill in the letter corresponding to each question above the question's answer in the decoder below.

1, 4, 2, 9, 5, 3

T: What is the range? ____________

L: What is the mean? ____________

I: What is the median? ____________

6, 7, 4, 8, 12, 2, 1, 2

U: What is the range? ____________

R: What is the mean? ____________

O: What is the median? ____________

E: What is the mode? ____________

What number in some data sets sticks out like a sore thumb?

___	___	___	___	___	___	___
5	11	8	4	3.5	2	5.25

Name ______________________ Date ______________ Class ______________

LESSON 7-3

Practice A

Bar Graphs and Histograms

The bar graph shows the lengths of four rivers. Use the graph for Exercises 1–3.

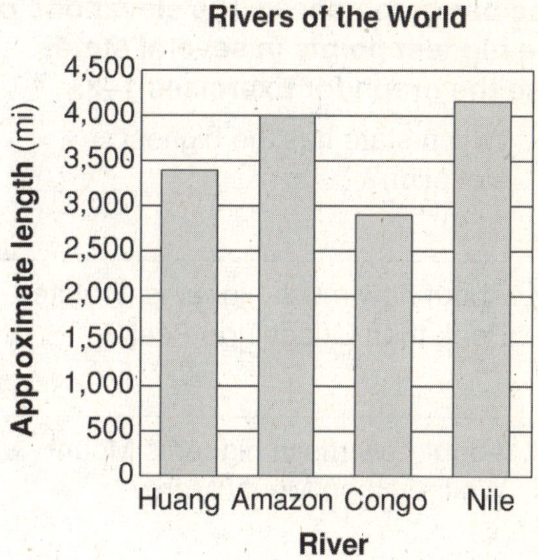

1. Which river is the longest?

____

2. About how much longer is the Amazon River than the Congo River?

____

3. About how much longer is the Nile River than the Huang River?

____

4. The table shows the ages of four U.S. Presidents when they first entered office. Make a bar graph of the data.

Name	President's Age
Truman	60
Kennedy	43
Carter	52
Bush	54

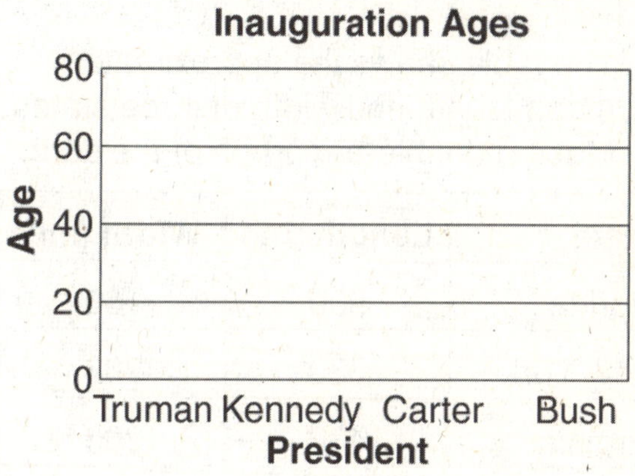

5. The table below shows the scores on a Spanish test. Make a histogram of the data.

Scores	Frequency
71–80	4
81–90	8
91–100	5

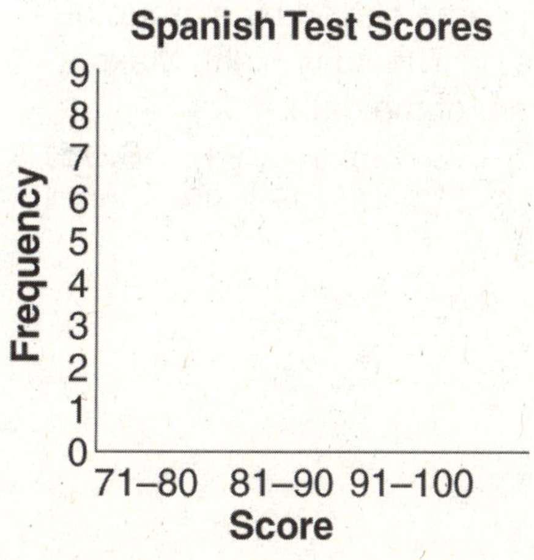

Name ______________________ Date ______________ Class ______________

LESSON 7-3

Practice B

Bar Graphs and Histograms

The bar graph shows the elevations of the highest points in several states. Use the graph for Exercises 1–3.

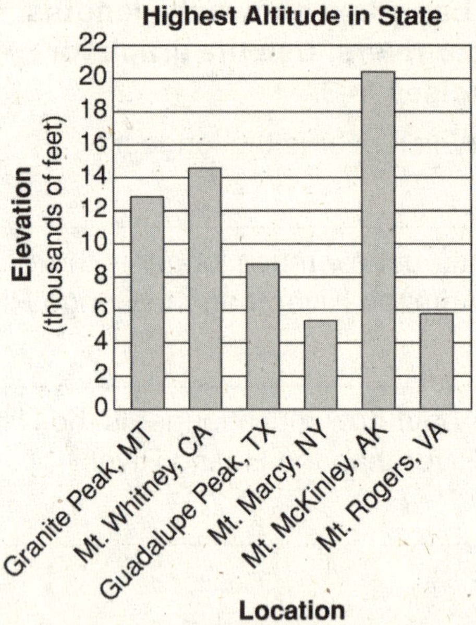

1. Which state has the highest elevation?

____

2. About how much higher is Granite Peak than Guadalupe Peak?

____

3. About how much higher is Mount Whitney than Mount Marcy?

____

4. The table shows the approximate mean length and width of three states. Make a double-bar graph of the data.

State	**Length** (mi)	**Width** (mi)
Florida	500	160
New York	330	283
Virginia	430	200

Key:

5. The list shows the bowling scores of the first game played by a group of bowlers on Thursday night. Make a histogram of the data.
96, 110, 132, 128, 105, 94, 116, 95, 126, 114, 123, 136, 121, 99

Name ______________________________ Date ______________ Class ______________

LESSON 7-3

Practice C

Bar Graphs and Histograms

The bar graph shows the activities in which most injuries occur for children ages 5–14. Use the graph for Exercises 1–3.

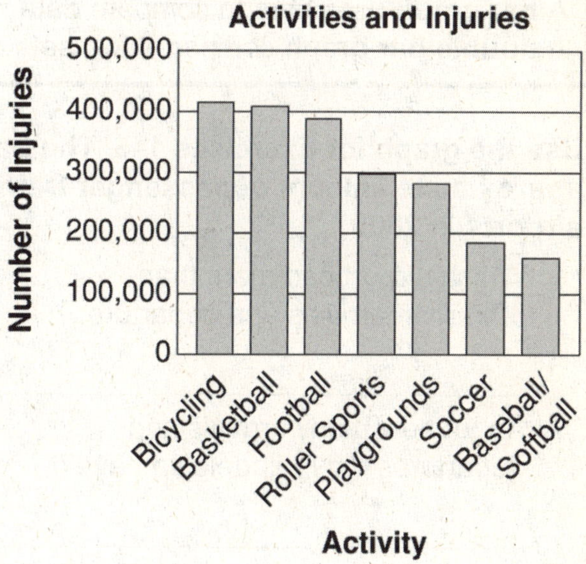

1. Which activity has the greatest number of injuries?

   ____

2. About how many more children are injured playing basketball than soccer?

   ____

3. About how many more children are injured participating in roller sports than on playgrounds?

   ____

4. The table shows the life expectancies of people living in three North American countries. Make a double-bar graph of the data.

Country	Male	Female
Canada	76	83
Greenland	66	74
Mexico	69	75

Key: ______________

5. The list shows the times, in minutes, in which runners finished a 10-kilometer race. Make a histogram of the data.

   92, 97, 100, 88, 85, 79, 76, 83, 92, 87, 78, 85, 79, 98, 83, 84, 86

Name ______________________ Date ____________ Class ____________

LESSON 7-3

Review for Mastery

Bar Graphs and Histograms

A **bar graph** uses bars to compare data.
A **double-bar graph** compares two sets of data.

Use the graph for Exercises 1–3. The bar graph shows the approximate amount of passenger traffic through some U.S. airports in 2005.

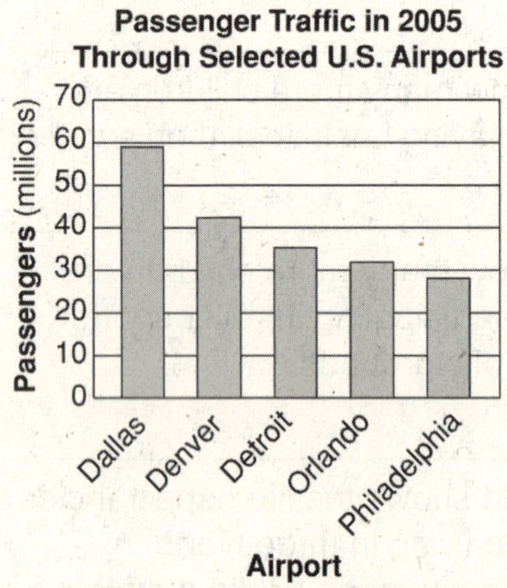

1. Which airport had more than 50 million arrivals and departures?

____

2. About how many arrivals and departures did Philadelphia have?

____

3. About how many more people passed through the Denver airport than the Orlando airport?

____

4. The table shows which sports are played by a seventh-grade class. Use the data in the table to make a double-bar graph.

Sport	Number of Boys	Number of Girls
Soccer	14	8
Tennis	7	10
Basketball	18	6
Swimming	12	20

a. Complete the vertical axis.

What is the vertical axis label?

____

What is the horizontal axis label?

____

b. Draw a bar on the horizontal axis for the number of boys playing each sport. Draw a bar for the number of girls playing each sport.

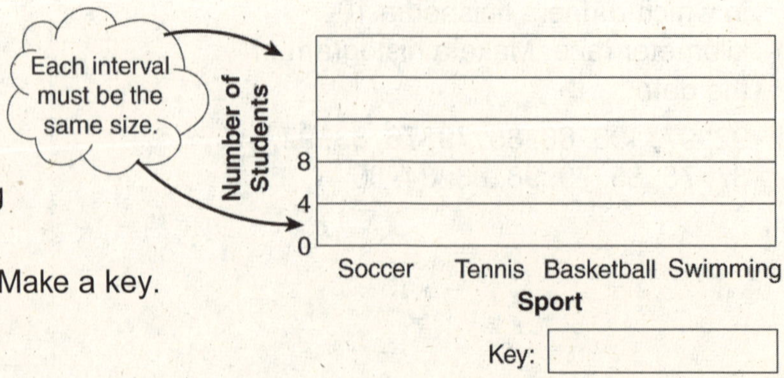

c. Give the graph a title. Make a key.

Name ______________________ Date ______________ Class ______________

LESSON 7-3

Review for Mastery

Bar Graphs and Histograms (continued)

A **histogram** is used to represent data in a frequency table. In a histogram the bars touch and both axes show numerical data.

Use the histogram for Exercises 5–7.

The histogram shows the pulse rates of 20 patients during check-ups.

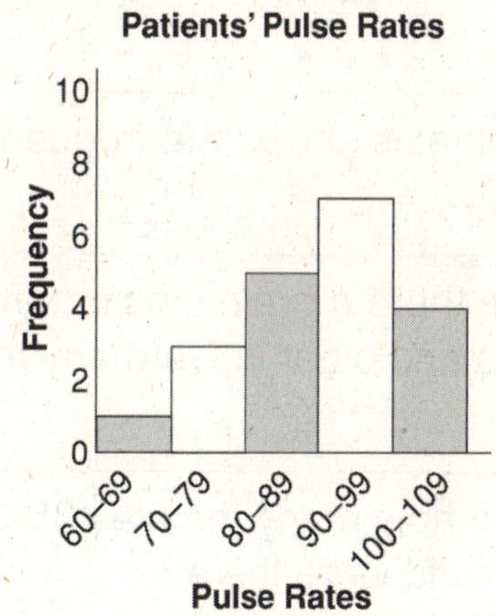

5. How many patients had a pulse rate between 80 and 89?

____

6. What range of pulse rates was most common among the patients?

____

7. How many more patients had a pulse rate between 100 and 109 than between 60 and 69?

____

8. The frequency table shows the length (in inches) of the fish caught on a class field trip to the lake. Use the data in the frequency table to make a histogram.

Fish Caught on Class Trip

Length (in.)	**Frequency**
1–5	1
6–10	6
11–15	8
16–20	5
21–25	2

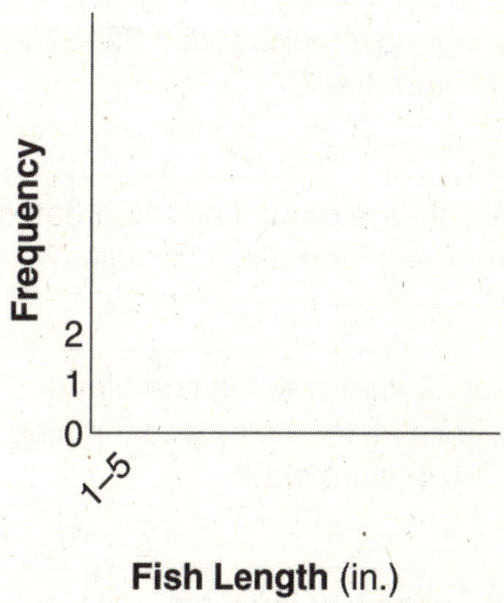

a. Complete the horizontal axis.

What is the vertical axis label? ______________________

What is the horizontal axis label? ______________________

b. The vertical axis shows the frequency. Finish labeling the axis.

c. Draw a bar for each interval. The bars should touch.

d. Give the graph a title.

Name ______________________ Date ______________ Class ______________

LESSON 7-3

Challenge

Sliding Histogram

The graph below is called a sliding histogram. It shows Germany's population in 2005.

Use the sliding histogram to the right to answer the questions.

1. What is the length of the interval used on the histogram?

____

2. Which axis shows the frequency?

____

3. Were there more males or females 80 years and older in Germany in 1997?

____

4. About how many females 35–39 years old were there?

____

5. About how many males 35–39 years old were there?

____

6. Which age group had the greatest number of females? of males?

____

Germany's Population 2005

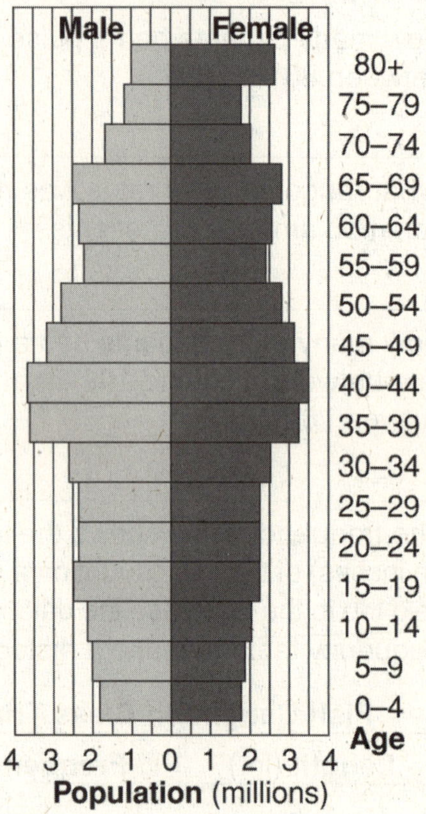

7. About what was the combined population of males and females 10–14 years old?

____

8. About how many more females 75–79 were there than males 75–79?

____

9. About how many more males were there 40–44 than 45–49?

____

10. At what age does the number of females start to differ significantly from the number of males?

____

Name ______________________ Date ______________ Class ______________

LESSON 7-3

Problem Solving

Bar Graphs and Histograms

Write the correct answer.

The double-bar graph shows the win-loss records for the Carolina Panthers football team in the years 1998–2004.

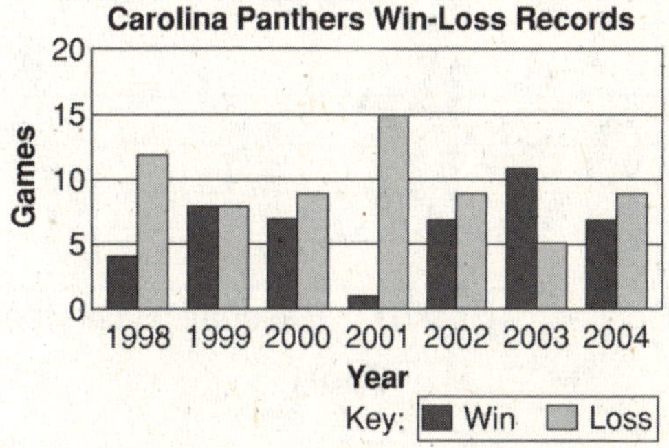

1. During how many seasons did the Panthers lose more games than they won?

____

2. In which year did the Panthers win more games than they lost?

____

3. Between which 2 years did the Panthers have the greatest improvement in their win-loss record?

____

4. In which year do you find the greatest range in the win-loss record?

____

Choose the letter for the best answer.

The histogram shows the ages of all members in a fan club.

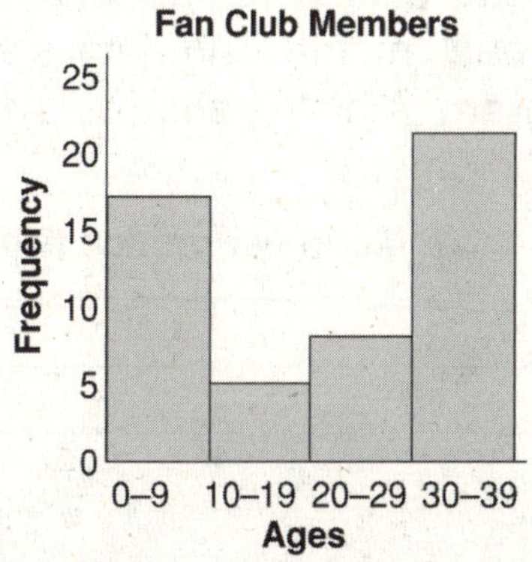

5. How many fan club members could be teenagers?

   A 5 C 17

   B 8 D 21

6. How many fan club members are between the ages of 30 and 39?

   F 5 H 17

   G 8 J 22

7. In which situation would you use a histogram to display data?

   A to show how you spend money

   B to show the change in temperature throughout the day

   C to show the golf scores from the whole team

   D to show the life expectancy of different animals

8. In which situation would you use a bar graph to display data?

   F to compare the speed of different computers

   G to show how a cat spends its time

   H to show how a child's height changes as he or she grows

   J to show the distribution of math grades in your class

Name ______________________ Date ______________ Class ______________

LESSON 7-3

Reading Strategies

Reading a Graph

Bar graphs create pictures for data.

A random sample was taken of boys' favorite sports.

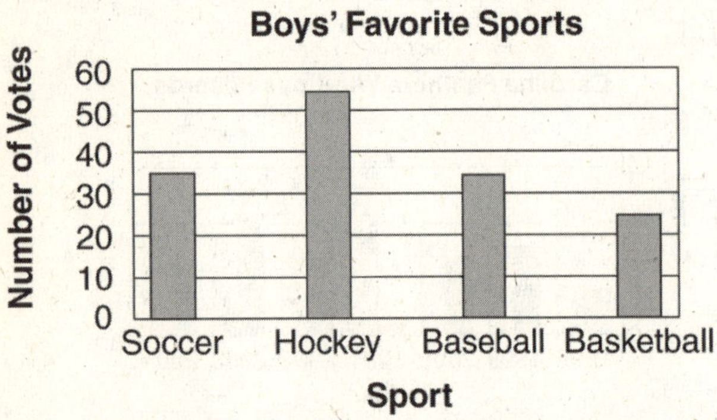

Answer the following questions about the bar graph.

1. What does the graph show? ______________________

2. What number does the scale of the graph count by? ______________________

3. How many boys like basketball best? ______________________

4. How many boys like hockey best? ______________________

Coaches were surveyed about the number of hours their teams practice each month. The data is shown in a frequency table and in a **histogram**. A histogram displays data in equal intervals with connected bars.

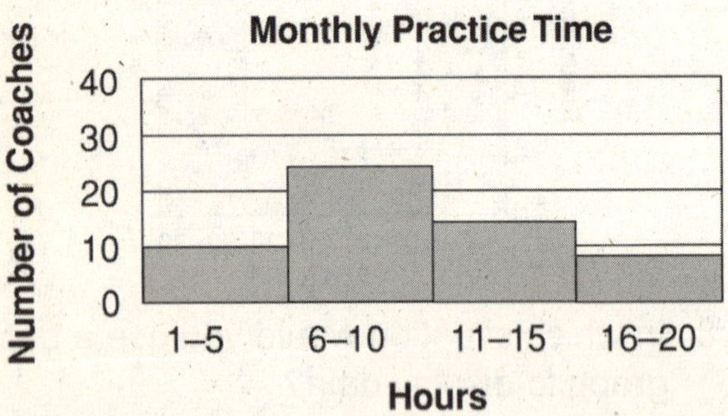

Answer the following questions about the histogram.

5. Where is the number of coaches located on the histogram?

____

6. The numbers on the left side of the histogram are in equal groups of what number?

____

7. What information is located along the bottom of the graph?

____

Name ______________________ Date ______________ Class ______________

LESSON 7-3

Puzzles, Twisters & Teasers

A Graphic Display!

Answer the question by studying the bar graph. Extract the letters from the graph to solve the riddle.

Number of Times Letters Appear in Answer to Riddle

Why was the bird scared to pretend to be sick?

It was performing an I __ __ – __ A G __ __ A __ T.

3 3 2 3 2 1

Name ______________________ Date ______________ Class ______________

LESSON 7-4

Practice A

Reading and Interpreting Circle Graphs

The circle graph directly below shows the results of a survey of 50 teens. They were asked about their favorite fruits. Use the graph for Exercises 1–3.

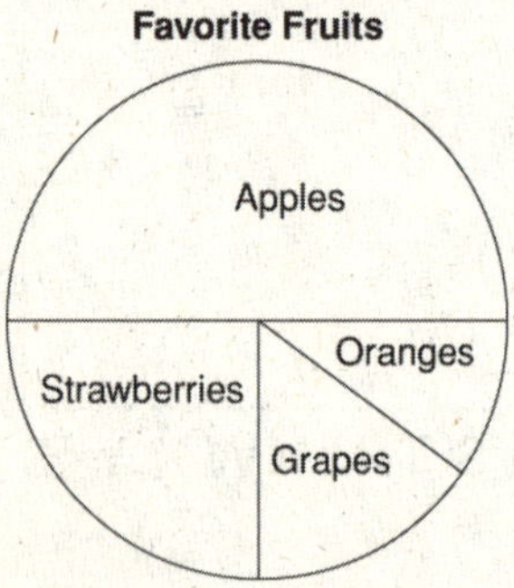

1. Did more teens pick apples or grapes?

____

2. About what percent of teens picked strawberries?

____

3. According to the survey, 10% of teens chose oranges. How many teens chose oranges?

____

The circle graph below shows the results of a survey of 100 people. They were asked about their favorite doughnut flavors. Use the graph for Exercises 4–6.

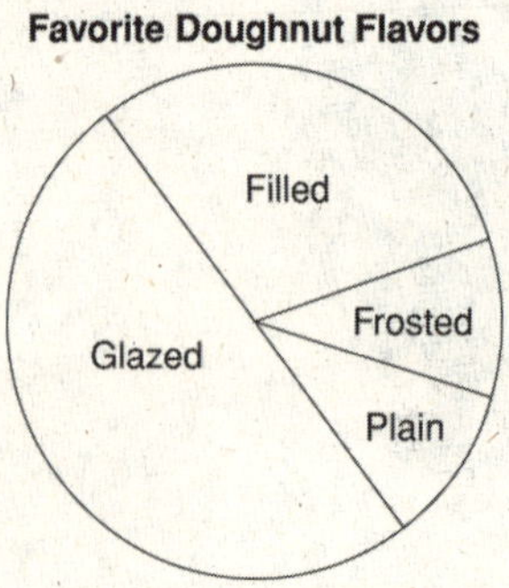

4. Did more people pick frosted or filled?

____

5. About what percent of people picked glazed?

____

6. According to the survey, 30% of the people chose filled doughnuts. How many people chose filled?

____

Would you use a bar graph or a circle graph to show the information? Explain your answer.

7. the number of bears that live in each of the national parks in Alaska

____

____

____

8. the number of visitors to Yellowstone National Park in August compared with the total number of summer visitors

____

____

____

Name ______________________ Date ______________ Class ______________

LESSON 7-4

Practice B

Reading and Interpreting Circle Graphs

The circle graph directly below shows the results of a survey of 80 teens who were asked about their favorite musical instruments. Use the graph for Exercises 1–3.

Favorite Musical Instruments

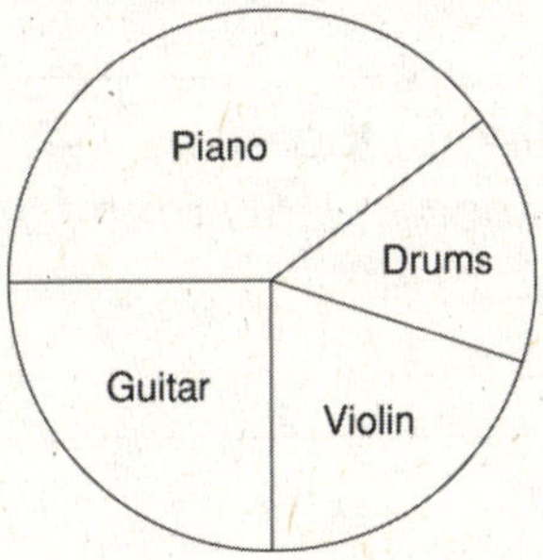

1. Did more teens pick piano or drums?

____

2. About what percent of teens picked guitar?

____

3. According to the survey, 20% of teens chose violin. How many teens chose violin?

____

The circle graph below shows the results of a survey of 100 people who were asked about their favorite vacation destinations. Use the graph for Exercises 4–6.

Favorite Vacation Destinations

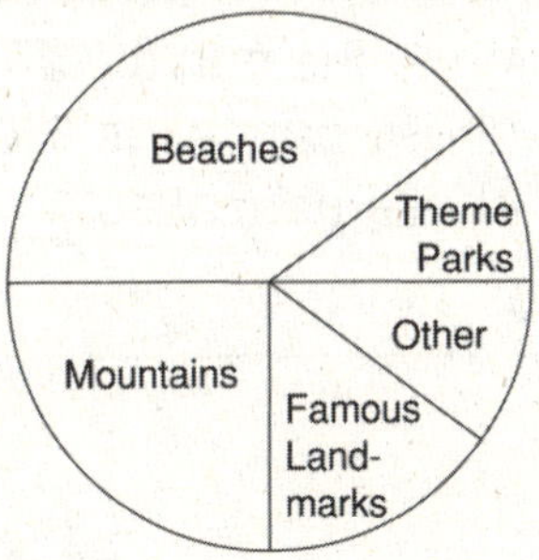

4. Did more people pick mountains or beaches?

____

5. About what percent of people picked mountains?

____

6. According to the survey, 15% of the people chose famous landmarks. How many people chose famous landmarks?

____

Decide whether a bar graph or a circle graph would best display the information. Explain your answer.

7. number of tornadoes in each state during one year

____

____

____

8. the number of pounds of Macintosh apples sold compared with the total number of pounds of apples sold at a market in one day

____

____

____

Name ____________________ Date ____________ Class ____________

LESSON 7-4

Practice C

Reading and Interpreting Circle Graphs

The circle graph directly below shows the results of a survey of 60 teens who were asked about their favorite sports. Use the graph for Exercises 1–3.

Favorite Sports

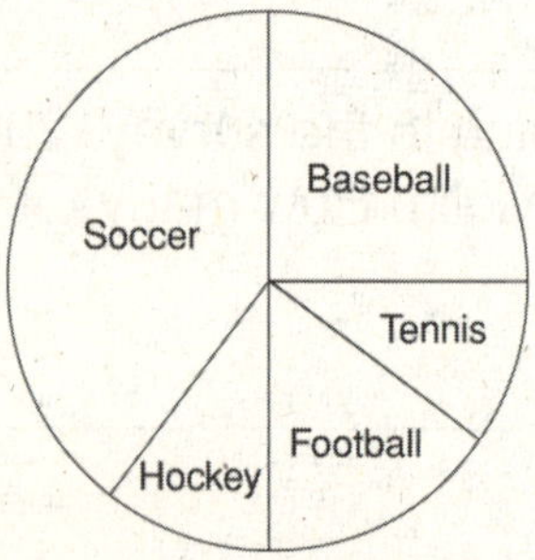

1. Did more teens pick baseball or hockey?

____

2. About what percent of teens picked tennis?

____

3. According to the survey, 40% of teens chose soccer. How many teens chose soccer?

____

The circle graph below shows the results of a survey of 200 people who were asked about their favorite breakfast foods. Use the graph for Exercises 4–6.

Favorite Breakfast Foods

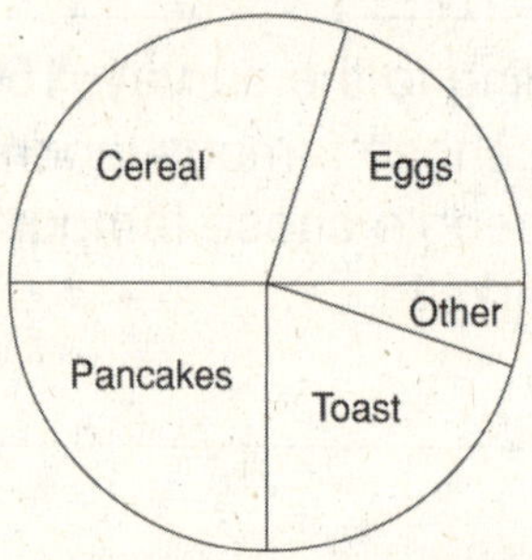

4. Did more people pick cereal or eggs?

____

5. About what percent of people picked toast?

____

6. According to the survey, 25% of the people chose pancakes. How many people chose pancakes?

____

Decide whether a bar graph or a circle graph would best display the information. Explain your answer.

7. the amount of money budgeted for entertainment in one month compared with the total monthly allowance

____

____

____

8. the number of calories different birds of prey need to consume in one day

____

____

____

Name ________________ Date ____________ Class ____________

LESSON 7-4

Review for Mastery

Reading and Interpreting Circle Graphs

A circle **graph** shows how the parts of a complete set of data are related. The circle shows 100% of the data.

Choose a circle graph to show what percent (or part) of the total sales is represented by each type of pizza.

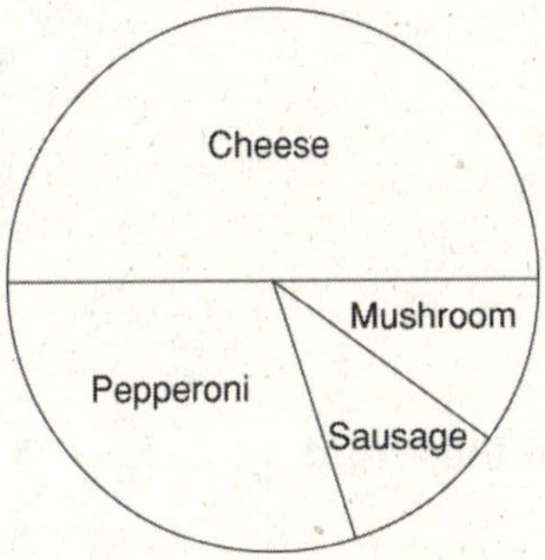

Choose a bar graph to show the number of each type of pizza sold.

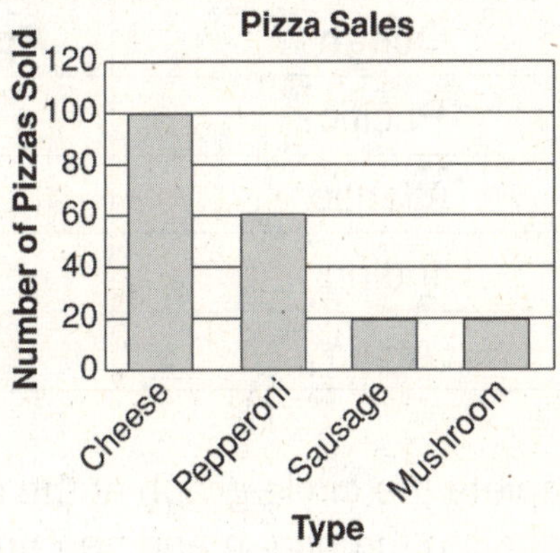

The circle graph shows that about one-half, or 50%, of the total sales was cheese pizzas. If 200 pizzas were sold, then about half, or 100, were cheese pizzas.

The circle graph shows the results of a survey of 100 teens who were asked about their favorite winter sport. Use the graph to answer each question.

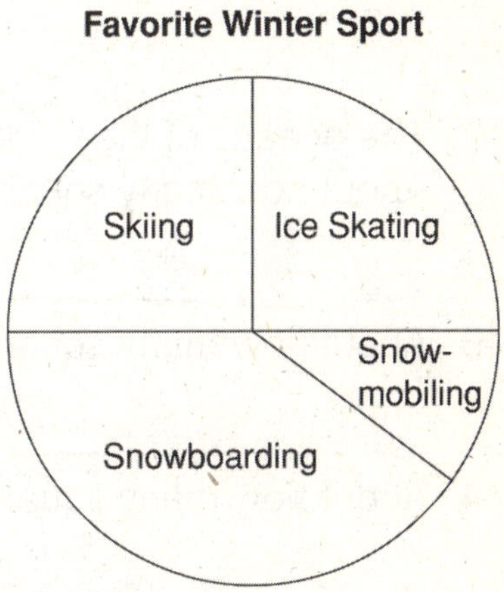

1. Did more teens pick skiing or snowboarding? ____________
2. About what percent of teens picked skiing? ____________
3. How many teens chose skiing? ____________

Decide whether a bar graph or a circle graph would best display the information. Explain your answer.

4. the number of hurricanes in each state during 1 year

____

____

____

5. the number of students who play soccer compared with the total number of students in a school

____

____

____

Name ______________________ Date ______________ Class ______________

LESSON 7-4

Challenge

Circle the Oceans

The table below shows the approximate percent of area that each ocean comprises of the total area of the oceans of the world. These oceans together actually make up one large, connected body of water.

Ocean	Approximate Area
Pacific	49%
Atlantic	25%
Indian	22%
Arctic	4%

1. Complete the circle graph at the right by labeling the ocean and percent of area for each part.

Oceans of the World

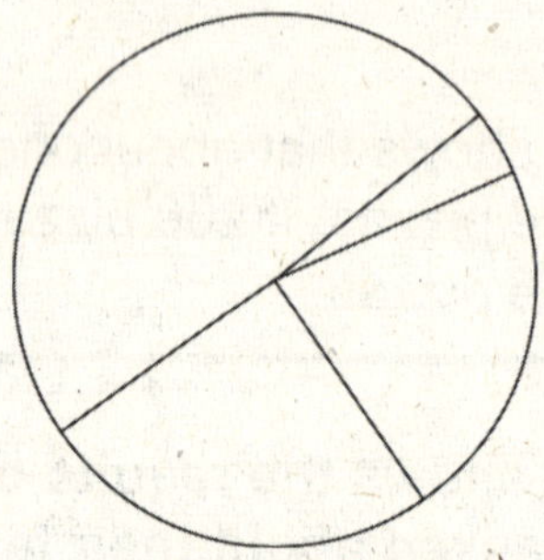

2. The oceans of the world cover almost 130 million square miles. About how many square miles are covered by the Pacific Ocean?

____

3. About how many square miles are covered by the Atlantic Ocean?

____

4. About how many square miles are covered by the Indian Ocean?

____

5. About how many square miles are covered by the Arctic Ocean?

____

6. About how many square miles are covered by the Pacific Ocean and the Atlantic Ocean combined?

____

7. About how many more total square miles are covered by the Atlantic, Indian, and Arctic oceans than by the Pacific Ocean?

____

Name ______________________ Date ______________ Class ______________

LESSON 7-4

Problem Solving

Reading and Interpreting Circle Graphs

Write the correct answer.

1. A market research group conducted a survey of 100 sports car owners. The group learned that 50% of the car owners loved their cars. What part of the circle in a circle graph would be represented by that statistic?

____

2. Juanita has 100 CDs. In her collection, 37 of the CDs are rock music, 25 are jazz, and 38 are country music. What part of the circle in a circle graph would represent the jazz CDs?

____

3. Mr. Martin wanted to compare his monthly rent to his total income. Should he use a circle graph or a bar graph?

____

4. Mr. Martin's rent has increased every year for the last 6 years. Should he use a circle graph or bar graph to show the yearly increase?

____

Choose the letter for the best answer. Use the circle graph.

Age of Fitness Club Members

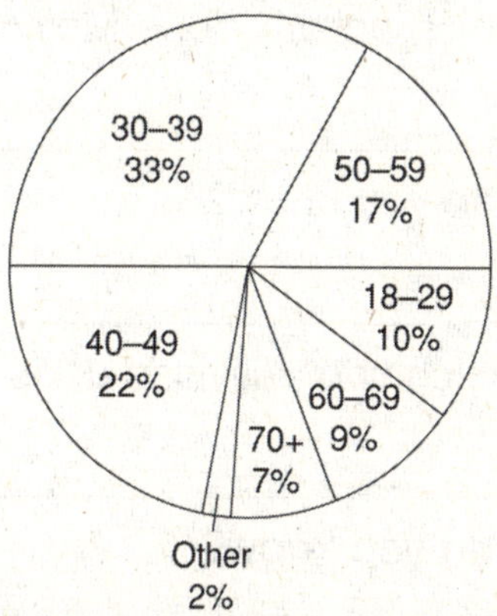

5. To which age group do most of the fitness club members belong?
   - A 18–20
   - B 70+
   - C 30–39
   - D 40–49

6. There are 100 members in a fitness club. How many members does the graph suggest will be between the ages of 18 and 39?
   - F 10
   - G 33
   - H 43
   - J 22

7. Which 2 age groups make up more than one-half the members?
   - A 18–29 and 30–39
   - B 30–39 and 40–49
   - C 40–49 and 50–59
   - D 18–29 and 70+

8. Which 2 age groups make up 3 times as many members as those who are between 60 and 69?
   - F 40–49 and 50–59
   - G 50–59 and 70+
   - H 30–39 and 18–29
   - J 18–29 and 50–59

Name ______________________ Date ______________ Class ______________

LESSON 7-4

Reading Strategies

Reading a Graph

This **circle graph** gives you the whole picture of Molly's 60-minute exercise program. You can see how each part of her exercise program relates to the whole program.

Molly's 60-Minute Exercise Program

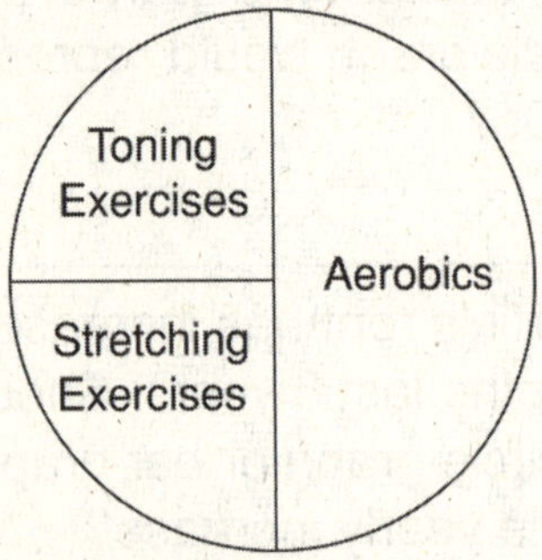

Answer the questions about the circle graph.

1. What does the graph show?

____

2. On which activity does Molly spend half of her time?

____

3. How do you know?

____

4. On which two activities does Molly spend the same amount of time?

____

5. How do you know?

____

You can figure out how many minutes Molly spends doing each activity by looking at the graph.

6. What is the total number of minutes Molly spends doing her full exercise program? ______________

7. How many minutes does Molly spend doing aerobics? (She spends one-half of her exercise program doing aerobics.) ______________

8. How many minutes does Molly spend doing stretching exercises? (She spends one-fourth of her time stretching.) ______________

Name ______________________ Date ______________ Class ______________

LESSON 7-4

Puzzles, Twisters & Teasers

Cow-abunga!

Doctor Digby did a study of people's eating habits. He asked 100 people what kinds of meat they prefer. Fifty people said they like all kinds of meat. Twenty-five people said they don't like any kind of meat. Five people said they prefer fish. Ten people said they prefer chicken. Ten people said they prefer beef. Complete the circle graph shown below. Use your answers to solve the riddle.

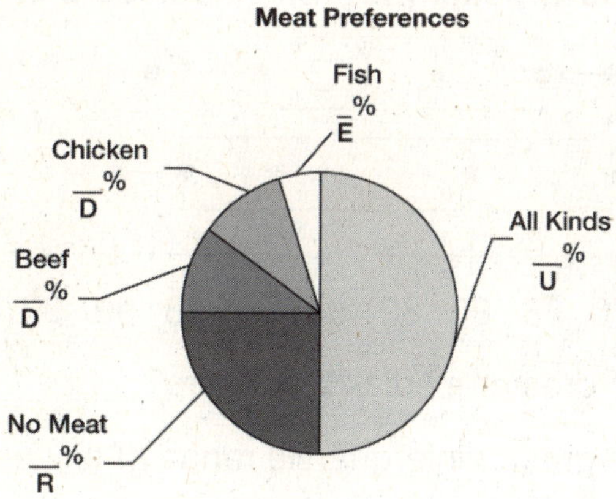

Why did the cow cross the road?

To get to the ______ ______ ______ ______ ______ side.

50% 10% 10% 5% 25%

Name ______________________ Date ______________ Class ______________

LESSON 7-5

Practice A

Box-and-Whisker Plots

1. Use the data to make a box-and-whisker plot.
   24, 32, 35, 18, 20, 36, 12

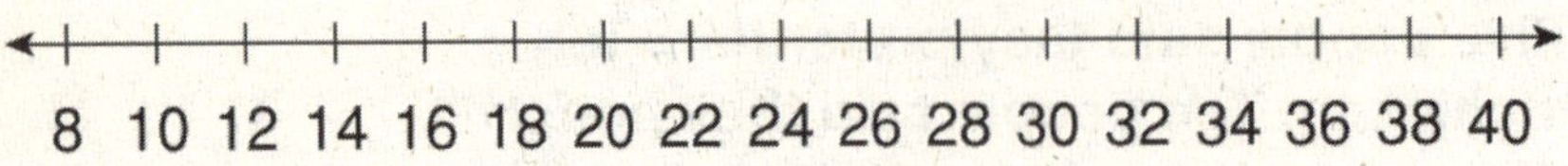

The box-and-whisker plot shows the test scores of two students. Use the box-and whisker plot for Exercises 2-4.

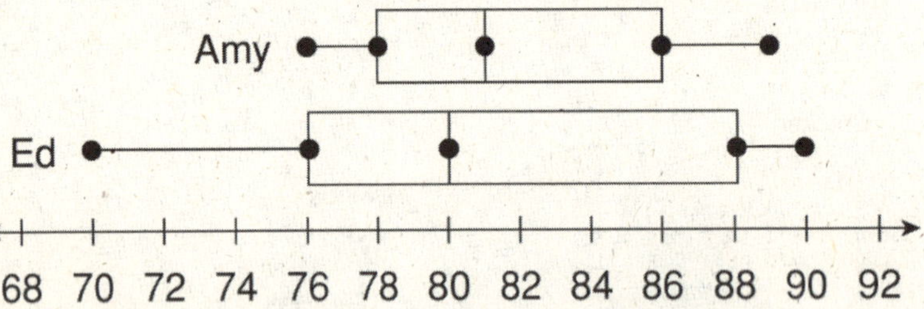

2. Which student has the greater median test score? ______________

3. Which student has the greater interquartile range of test scores? ______________

4. Which student has the greater range of test scores? ______________

5. Which student appears to have more predictable test scores? Explain your answer.

____

____

The box-and-whisker plot shows prices of hotel rooms in two beach towns. Use the box-and whisker plot for Exercises 6-8.

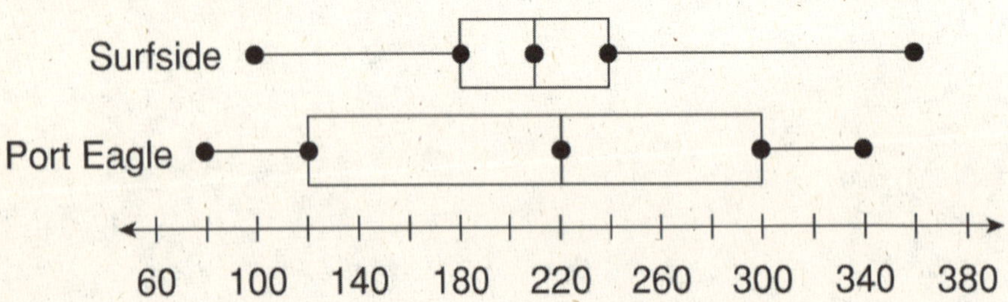

6. Which town has the greater median room price? ______________

7. Which town has the greater interquartile range of room prices? ______________

8. Which town appears to have more predictable room prices? Explain your answer.

____

____

Name ______________________ Date ______________ Class ______________

LESSON 7-5

Practice B

Box-and-Whisker Plots

1. Use the data to make a box-and-whisker plot.
   19, 46, 37, 16, 24, 47, 23, 19, 31, 25, 42

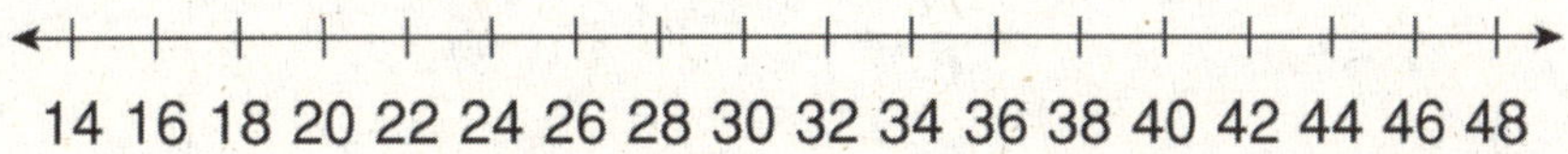

Use the box-and-whisker plot of games won per season by the New York Yankees and the Arizona Diamondbacks for 1998–2005 for Exercises 2–4.

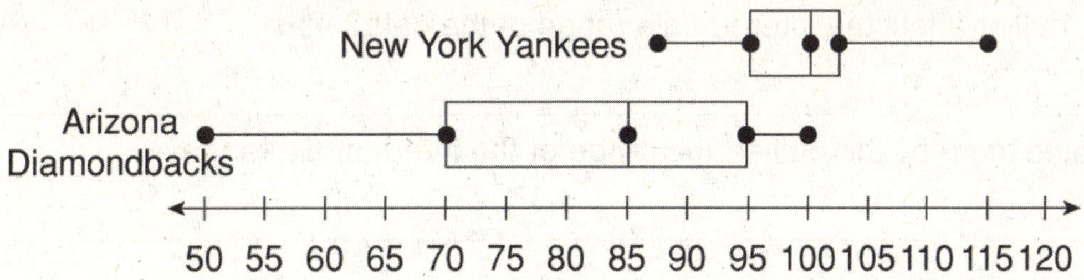

2. Which team has the greater median number of games won? ______________

3. Which team has the greater interquartile range of games won? ______________

4. Which team appears to have a more predictable performance? ______________

Use the box-and-whisker plot of nightly tip totals that a waitress gets at two different restaurants for Exercises 5–7.

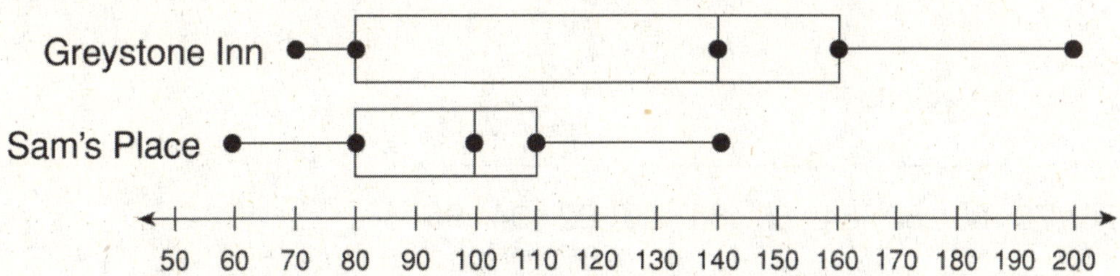

5. At which restaurant is the median tip total greater? ______________

6. At which restaurant is the interquartile range of tip totals greater?

   ____

7. At which restaurant does the tip total appear to be more predictable?

   ____

Name ________________________ Date ______________ Class ______________

LESSON 7-5

Practice C

Box-and-Whisker Plots

Use the data for Exercises 1–3.

38, 42, 26, 32, 40, 28, 36, 27, 29, 6, 30

1. Make two box-and-whisker plots of the data on the same number line: one plot with the outlier and one plot without the outlier.

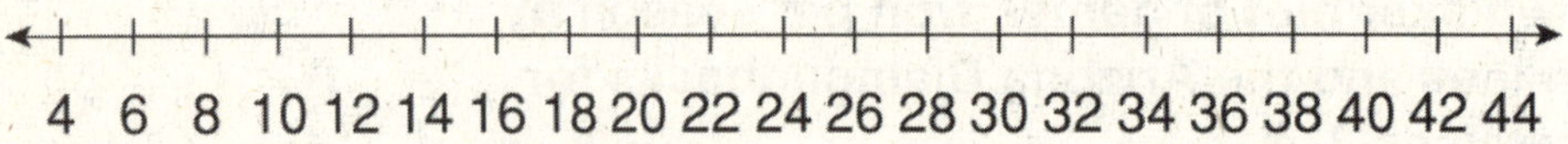

2. How does the outlier affect the interquartile range of the data?

____

3. Which is affected more by the outlier: the range or the interquartile range?

____

The table shows scores for two golfers. Use the table for Exercises 4–7.

Henry	78	80	74	91	73	88	92	94	78	80
Trish	82	84	81	82	80	89	86	90	78	85

4. Make two box-and-whisker plots of the data on the same number line.

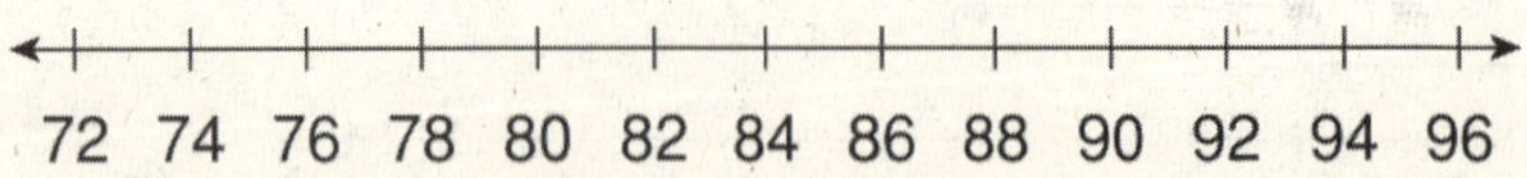

5. Which golfer has the lower median score? ______________

6. Which golfer has the lesser interquartile range of scores? ______________

7. Which golfer appears to be more consistent? ______________

Name ______________________ Date ______________ Class ______________

LESSON 7-5

Review for Mastery

Box-and-Whisker Plots

A **box-and-whisker plot** separates a set of data into four equal parts.

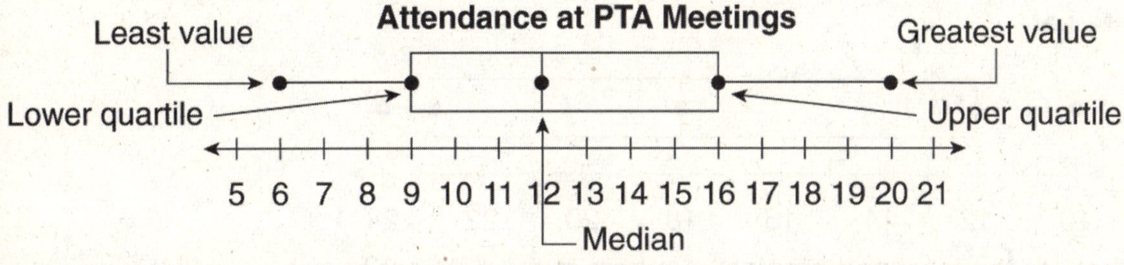

Use the data to create a box-and-whisker plot on the number line below:
35, 24, 25, 38, 31, 20, 27

1. Order the data from least to greatest.

   ____

2. Find the least value, the greatest value, and the median.

   ____

3. The **lower quartile** is the median of the lower half of the data. The **upper quartile** is the median of the upper half of the data. Find the lower quartile and the upper quartile in this data.

   ____

4. Above the number line, plot points for the least value, lower quartile, median, upper quartile, and greatest value. Draw a box around the quartiles and the median. Draw a whisker from the least value to the lower quartile. Draw a whisker from the upper quartile to the greatest value.

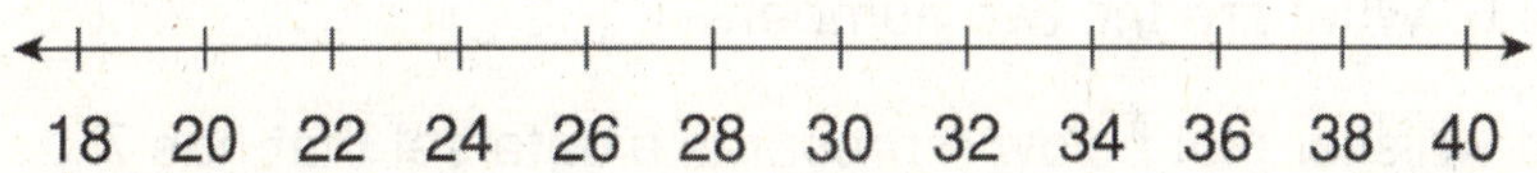

Use the data to create a box-and-whisker plot:
63, 69, 61, 74, 78, 72, 68, 70, 65

5. Order the data. ______________________

6. Find the least and greatest values. ______________

7. Find the median, lower quartile, and upper quartile. ______________

8. Plot the points, draw the box, and add the whiskers.

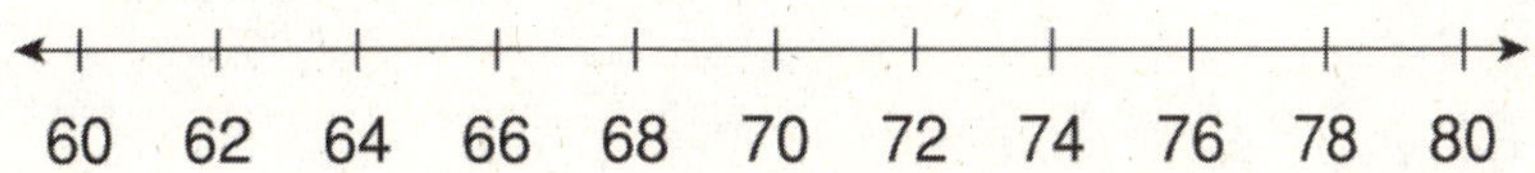

Name ______________________ Date ______________ Class ______________

LESSON 7-5

Challenge

Puzzling Plots

The box-and-whisker plot below represents a data set that has 7 numbers. Use the box-and-whisker plot to find the numbers in the data set.

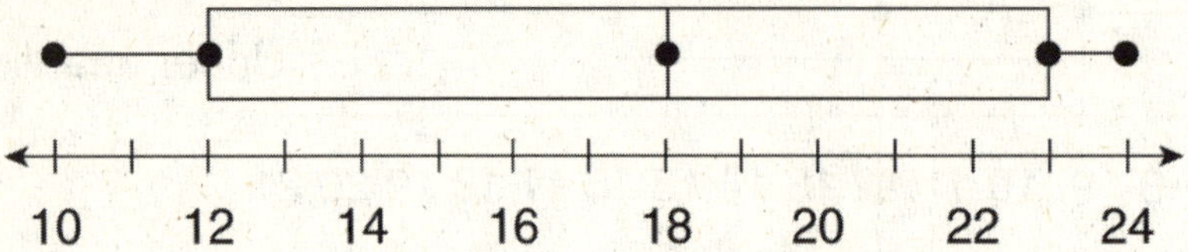

1. What are the least and greatest numbers in the data set? ______________

2. What is the median? ______________

3. Is the median one of the 7 numbers in the data set, or is it the mean of the middle two numbers. Explain your answer.

____

____

4. What are the lower and upper quartiles? ______________

5. Are the lower and upper quartiles two of the numbers in the data set? Explain your answer.

____

____

6. The remaining two numbers in the data set have a mean of 19 and range of 6. What are the two numbers? ______________

7. The box-and-whisker plot below represents a data set that has 7 numbers. When the numbers are in order from least to greatest, the third and fifth numbers in the set have a mean of 12 and a range of 4. What are the 7 numbers?

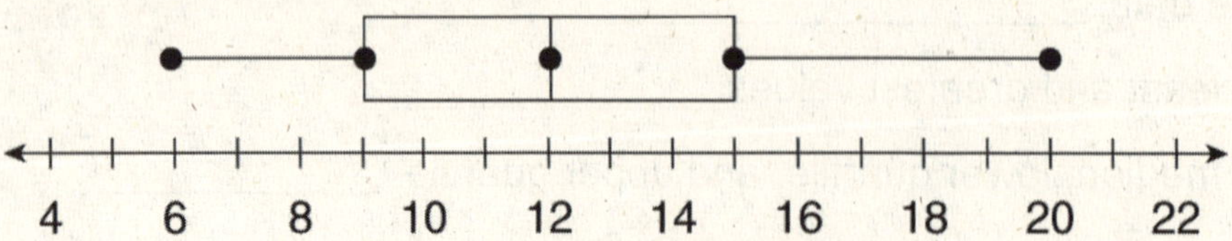

____

Name ______________________ Date ______________ Class ______________

LESSON 7-5

Problem Solving

Box-and-Whisker Plots

Write the correct answer.

A fitness center offers two different yoga classes. The attendance for each class for 12 sessions is represented in the box-and-whisker plot.

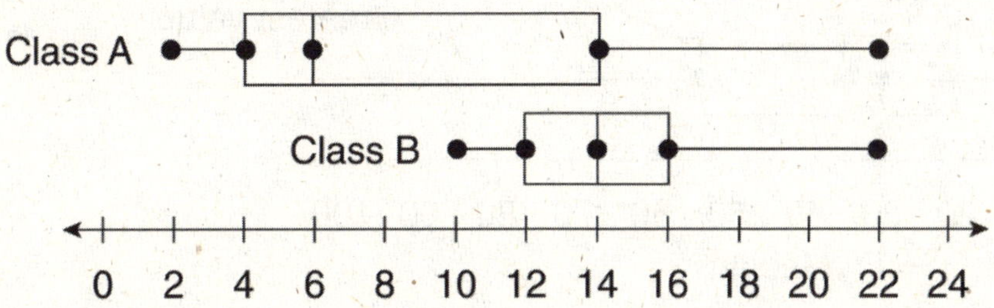

1. Which class has a greater median attendance? How much greater is it?

____

2. Which class appears to have a more predictable attendance?

____

3. Which class has an attendance of less than 14 people 75% of the time?

____

4. What percent of the time does Class B have an attendance greater than 16?

____

Choose the letter for the best answer.

The box-and-whisker plot shows the percent of people in eight Central American countries who used the Internet in 2005.

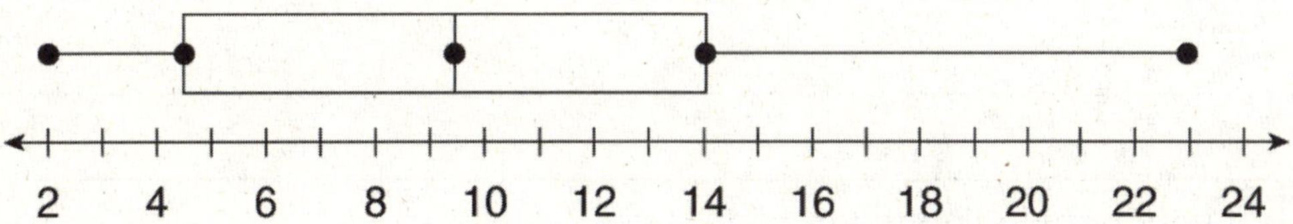

5. What is the range in the percents of people who used the Internet in the eight countries?

   A 23%
   B 21%
   C 14%
   D 9.5%

6. In how many of these eight countries did at least 9.5% of the people use the Internet?

   F less than 2
   G less than 4
   H at least 4
   J cannot be determined

7. What is the mode of the data?

   A 2%
   B 4.5
   C 9.5%
   D cannot be determined

8. What is the interquartile range of the data?

   F 23%
   G 21%
   H 14%
   J 9.5%

Name ______________________ Date ______________ Class ______________

LESSON 7-5

Reading Strategies

Use Graphic Aids

A **box-and-whisker plot** shows a set of data divided into four equal parts called **quartiles.**

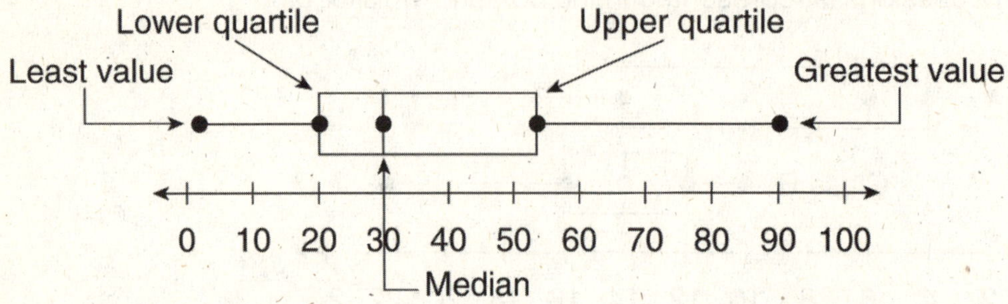

- The median score divides the set of data in half. The median score for this plot is 30.
- The box shows the middle half of the data, located on either side of the median. The box extends from 20 to 53.
- The two whiskers identify the remaining half of the data. One whisker extends from the box to the greatest value: from 53 to 90. The other whisker extends from the box to the least value: from 2 to 20.

Answer each question.

1. What does the box stand for in a box-and-whisker plot?

____

2. How are the whiskers determined?

____

____

3. Why is it important to find the median score?

____

Describe where these scores are located in the box-and-whisker plot above.

4. 18 between the ____________________ and the ____________________.

5. 75 between the ____________________ and the ____________________.

6. 45 between the ____________________ and the ____________________.

Name ______________________ Date ______________ Class ______________

LESSON 7-5

Puzzles, Twisters & Teasers

The Plot Thickens!

Circle words from the word list in the word search below.

box	whisker	plot	first	third	upper
quartile	extreme	data	median	range	fraction

E	B	O	X	Z	D	W	P	Y	J	I	O	R	C
X	F	D	Q	O	W	E	R	F	C	O	D	A	T
P	R	R	U	E	H	H	T	I	P	J	A	D	A
L	A	D	A	W	T	H	I	R	D	U	T	B	M
O	C	A	R	C	S	B	I	S	R	P	A	I	E
T	T	O	T	A	T	A	L	T	K	P	G	H	D
U	I	K	I	O	N	M	T	N	F	E	A	Z	I
P	O	M	L	D	E	G	R	D	G	R	R	W	A
L	N	B	E	X	T	R	E	M	E	E	S	V	N

Now think about the alphabet and the letters you see above.

What two letters of the alphabet hold nothing when you say them together?

__________ __________

CHAPTER 7

Family Letter

7B Representing and Analyzing Data

Dear Family,

The student is learning to display and analyze data that changes over a period of time in a line graph or a double-line graph.

The table shows the number of cars sold at AZ Cars during one year. Make a line graph to determine between which months there was the greatest increase in sales.

Jan	Feb	Mar	Apr	May	Jun	Jul	Aug	Sep	Oct	Nov	Dec
25	5	32	38	43	20	13	15	9	11	30	28

Step 1 Determine the scale and the interval for each axis.
Step 2 Plot a point for each pair of values.
Step 3 Connect the points using line segments.
Step 4 Label the axes and give the graph a title.

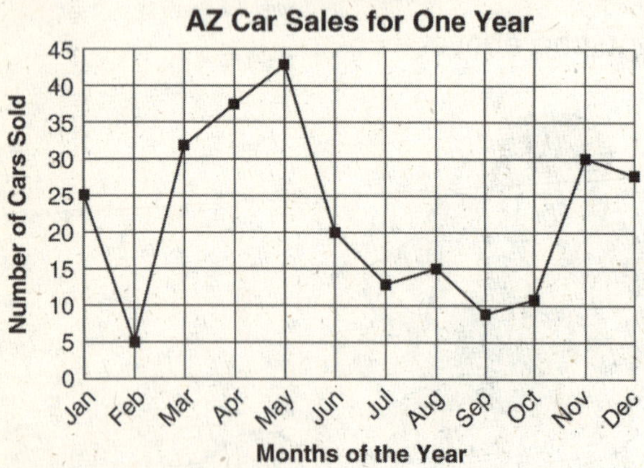

AZ Cars had the greatest increase between February and March.

Another way to display sets of data is a **Venn diagram.** A Venn diagram consists of overlapping circles. Each circle represents a set of data. The overlapping area represents the number of items that share the same traits.

Pets in Sarah's Class

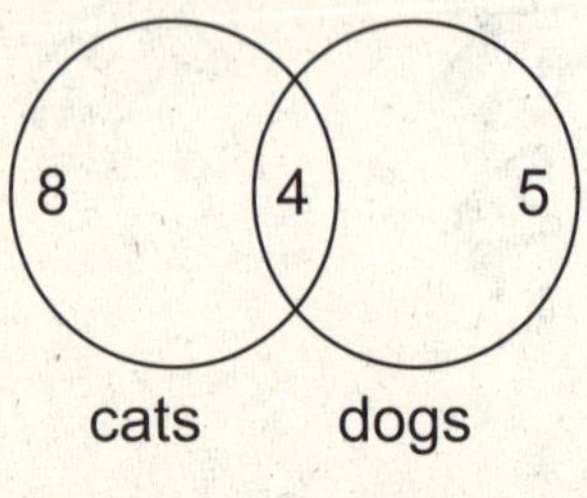

Pets in Sarah's Class

Cat	Dog	Both
𝍸 𝍸 ||	𝍸 ||||	||||

17 people in Sarah's class have pets. 8 have a cat, 5 have a dog, and 4 have both a cat and a dog.

Vocabulary

These are the math words we are learning:

biased sample a group that does not fairly represent the population

convenience sample a group of the population that is readily available

correlation the description of the relationship between two data sets

double-line graph a graph that shows change over time for two sets of data

line graph a graph that uses line segments to connect data points

negative correlation the values in one data set increase as the values in the other data set decrease

no correlation the values in both data sets show no pattern

population a whole group that is to be studied

positive correlation the values in both data sets increase at the same time

random sample a representative sample that gives every member of the population an equal chance of being chosen

CHAPTER 7

Family Letter

7B Representing and Analyzing Data continued

Sometimes data is gathered to find out if a relationship between two different groups exists. The student will learn how to create a scatter plot and determine whether a correlation between the sets of data exists.

Use the data to make a scatter plot. Describe the relationship between the data sets.

Step 1 Plot the points.

Step 2 Determine whether a relationship exists.

Weight (in lb)	Time Spent Exercising (in hours per week)
145	2
175	1
125	4
168	3
132	1
105	5
150	2

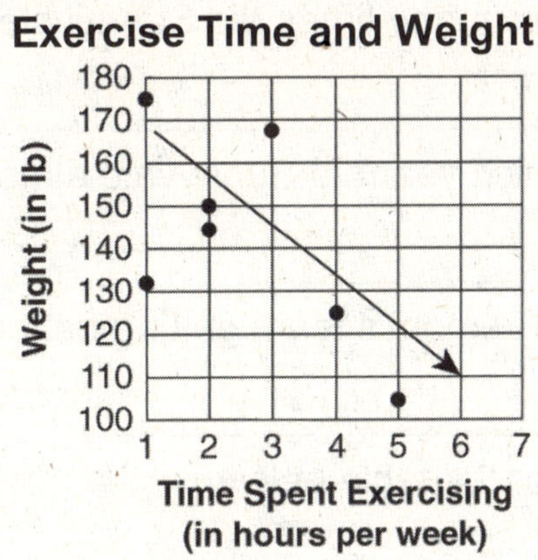

There seems to be a **negative correlation** between how much time a person spends exercising and his or her weight.

The student has learned many different ways to display data. He or she will also learn how to choose the best method to display different kinds of data. For example, if someone wants to know the range of the data, a box-and-whisker plot will be helpful.

It is important that the student learn to distinguish an accurate portrayal of data versus a misleading one. Look for real-world data displays with the student. Check to see whether these displays are misleading.

Sincerely,

sample a part of a group or population that is being studied

scatter plot a display of ordered pairs of data on a coordinate plane that shows how two variables relate to each other

Name ______________________ Date ______________ Class ______________

CHAPTER 7

At-Home Practice

7B Representing and Analyzing Data

Use the table for Exercises 1–4.

1. Make a double-line graph to show the data from the table. Be sure to label the axes and title the graph.

Basketball Scores

Year	2003	2004	2005	2006	2007	2008
Fremont	56	85	99	115	123	125
Sunnyvale	132	152	144	125	101	129

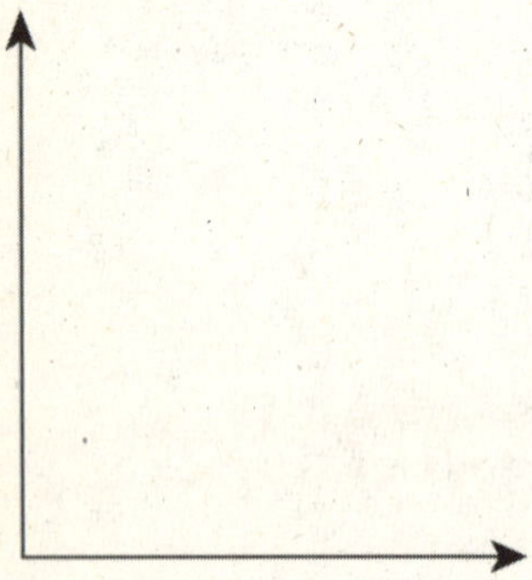

2. How many times did Sunnyvale beat Fremont?

____

3. Which year did Sunnyvale almost double Fremont's score?

____

4. Is Fremont's team getting better or worse? How do you know?

____

5. Make a Venn diagram for the table below.

Mrs. Hill's Music Students

Piano Lessons	**Singing Lessons**	**Both**
13	17	5

Choose the most appropriate way to display each set of data.

6. The relationship between age and height

____

7. Favorite sports of a random sample by percent

____

8. Bank account balance over a period of time

____

Answers: **1.** Graph should have years on the *x*-axis and scores on the *y*-axis. **2.** 5 **3.** 2004 **4.** Better, because each year they score more points. **5. Mrs. Hill's Music Students** **6.** scatter plot **7.** circle graph **8.** line graph

Name ______________________ Date ____________ Class ____________

CHAPTER 7

Family Fun

7B Scatter Plots!

Directions

Your job is to determine whether a relationship exists between two very different variables.

- Choose which set of variables you are going to use from the list below or make up your own pair of curious variables.
    - Shoe Size versus Waist Measurement
    - Hair Color versus Grade in Math Class
    - Eye Color versus Height
    - Height versus Amount of Water Consumed in a Day
    - Leg Length versus Amount of Time Spent Watching TV
- Interview your friends and family to obtain the data. Use the table below to record your data.
- Plot your points on the graph. Be sure to label your axes and title your graph.
- Use the scatter plot to determine what type of correlation exists between your two variables: positive correlation, negative correlation, or no correlation.

Were you surprised by the results? ______________________

Why or why not? ______________________

My graph has a ______________________ correlation between these variables.

Name ______________________ Date ______________ Class ______________

LESSON 7-6

Practice A

Line Graphs

Use the table to answer the questions.

U.S. Personal Spending on Selected Electronics	
Year	**Amount Spent** (\$billions, estimated)
1994	\$71
1996	\$80
1998	\$90
2000	\$107

1. Use the data in the table to make a line graph.

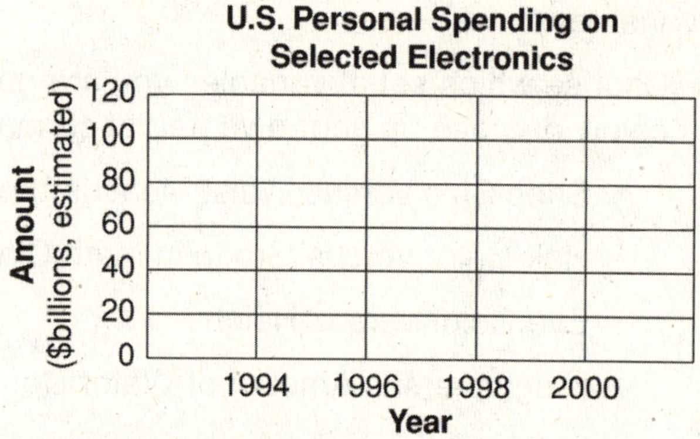

2. When did the amount spent on electronics increase the most?

____

3. About how much was spent on electronics in 1999?

____

The table below shows normal monthly temperatures in Asheville, North Carolina, and Miami, Florida, for the first 4 months of the year. Use the table to answer the questions.

	Jan.	Feb.	Mar.	Apr.
Asheville	36 °F	39 °F	47 °F	55 °F
Miami	67 °F	69 °F	72 °F	75 °F

4. Make a double-line graph of the data.

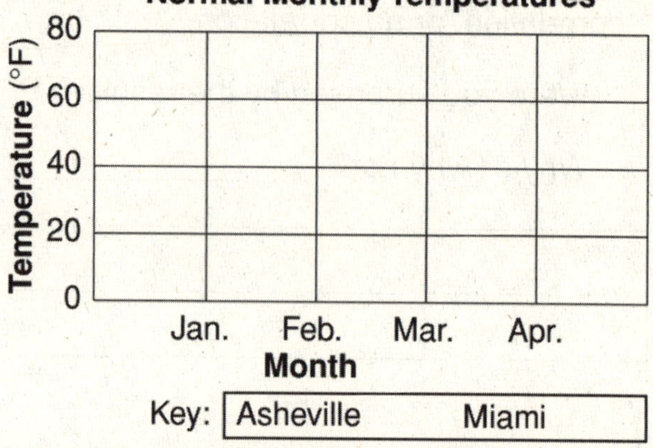

5. During which month is there the greatest difference in temperatures between the two cities?

____

6. During which month is there the least difference in temperatures between the two cities?

____

7. In which city did the temperature vary the least during these 4 months?

____

8. Where should you spend your April vacation if you want to go hiking wearing your new sweatshirt?

____

Name ______________________ Date ______________ Class ______________

LESSON 7-6

Practice B

Line Graphs

Use the table for Exercises 1–3.

Retail Price of Regular Gasoline in the United States (to the nearest cent)

Year	1990	1992	1994	1996	1998	2000	2002	2004
Price Per Gallon	\$1.16	\$1.13	\$1.11	\$1.23	\$1.06	\$1.54	\$1.32	\$1.82

1. Make a line graph of the data.

2. When did the cost of gasoline decrease the most?

____

3. About how much did gasoline cost in 1995?

____

The table below shows the student population at elementary schools in two cities, New City and Jackson.

Year	1996	1997	1998	1999	2000	2001	2002	2003
New City	450	460	440	430	495	500	600	645
Jackson	500	475	450	525	430	440	485	480

4. Make a double-line graph of the data.

Key:

5. During which year did New City's school population increase the most?

____

6. The mall in Jackson closed. Many people lost their jobs and moved their families to New City, where a new mall opened. In what year did this probably happen? Explain your thinking.

____

____

____

Name ______________________ Date ____________ Class ____________

LESSON 7-6

Practice C

Line Graphs

Use the table for Exercises 1–3.

U.S. Population (to the nearest million)	
Year	**Number**
1820	10
1840	17
1860	31
1880	50
1900	76
1920	106
1940	132
1960	179
1980	226
2000	281

1. Make a line graph of the data.

2. When did the U.S. population have the greatest increase?

____

3. Estimate the U.S. population in 1970.

____

The table below shows the price in cents per pound U.S. farmers received for apples and peanuts they grew.

Year	1996	1997	1998	1999	2000	2001	2002	2003
Apples (cents/lb)	20.8	22.1	17.1	21.3	17.8	22.9	25.6	29.5
Peanuts (cents/lb)	28.1	28.3	25.7	25.4	27.4	23.4	18.2	18.8

4. Make a double-line graph of the data in the table.

Key:

5. During which year was there the greatest difference in prices?

____

6. During which year was there the least difference in prices?

____

7. What trends do you see in the data?

____

____

____

Name ______________________ Date ______________ Class ______________

LESSON 7-6

Review for Mastery

Line Graphs

A **line graph** shows data changing over time.

A **double-line graph** shows two sets of data changing over time.

1. The table shows the normal daily low temperature in °F in Indianapolis and San Francisco. Use the data to make a double-line graph.

Month	Jan.	Mar.	May	Jul.	Sept.	Nov.
Indianapolis	26°	41°	63°	75°	67°	43°
San Francisco	49°	53°	58°	63°	65°	55°

a. Finish labeling each axis.

What is the vertical axis label?

____

What is the horizontal axis label?

____

b. Plot each point for each set of data.

c. Connect the points. Use a dotted line to represent one set of data.

d. Give the graph a title and key.

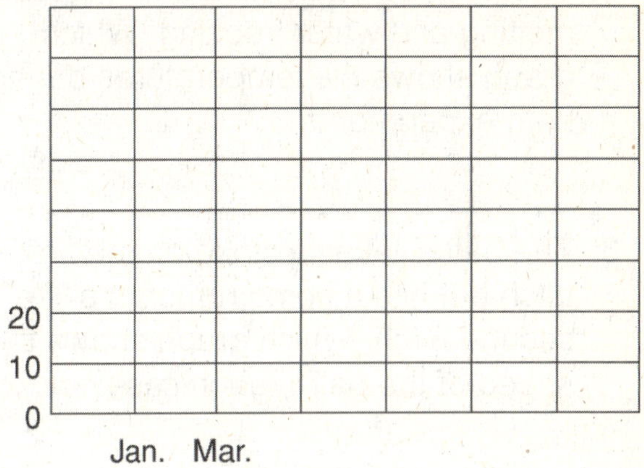

Key: ______________________

2. In which city was the temperature the most constant?

____

3. During which 2 months in Indianapolis did the temperature increase the most?

____

4. Use the graph to estimate the low temperature in October in San Francisco.

____

5. During which month was the difference in the daily normal low temperatures in Indianapolis and San Francisco most likely the smallest?

____

Name ______________________ Date ______________ Class ______________

LESSON 7-6

Challenge

What's My Line?

Write the letter of the graph that best represents each event described.

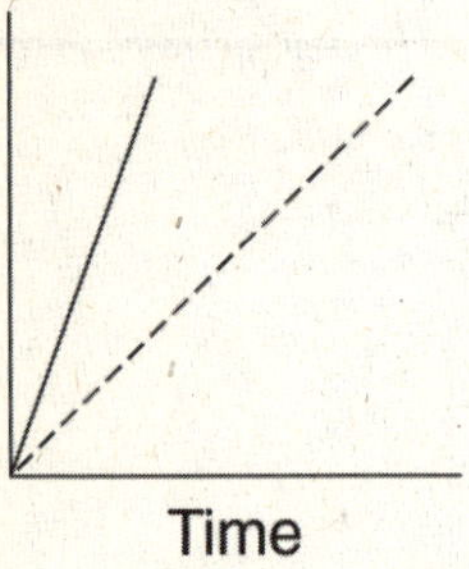

A

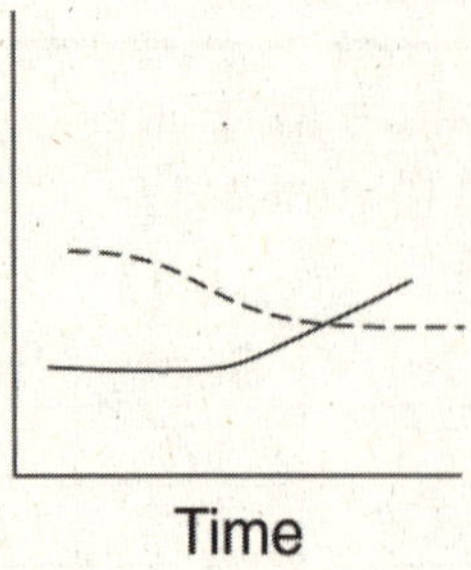

B

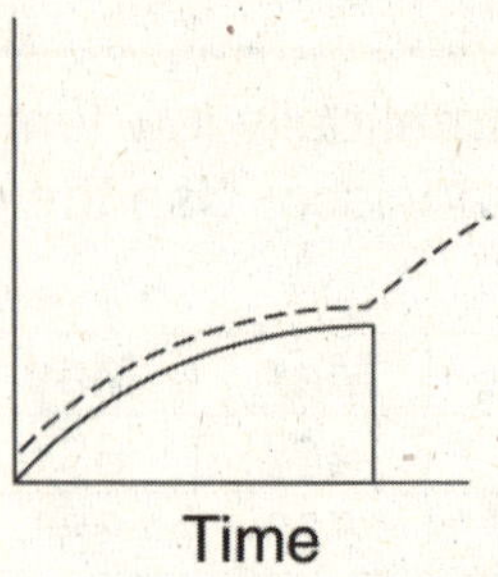

C

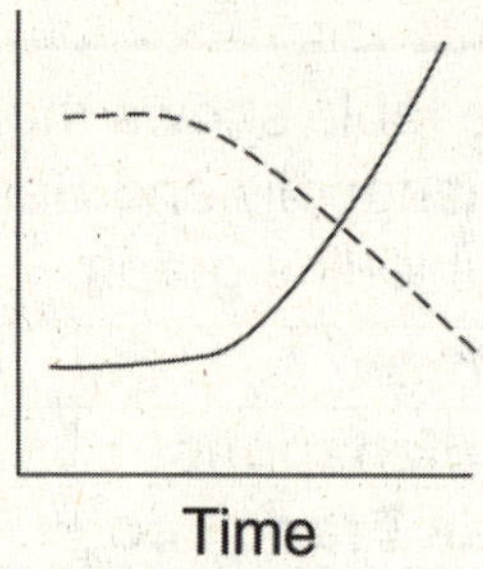

D

1. Two science students are timing ice melting and water freezing. Which graph shows the temperatures during both experiments?

   ____

2. In the music industry, when CDs became popular, sales of cassettes plummeted. Which graph shows this trend?

   ____

3. The batter takes a strike on the first pitch but hits a home run on the second pitch. Which graph shows the speed of the ball in each case?

   ____

4. Two brothers attend the same school. One walks and the other rides a bike. Which graph compares the time required for them to travel to school?

   ____

Draw a double-line graph to solve.

5. The town swimming pool puts aside the last 15 minutes of every hour for an adult swim. Show the change in the number of children and adults in the pool over several hours.

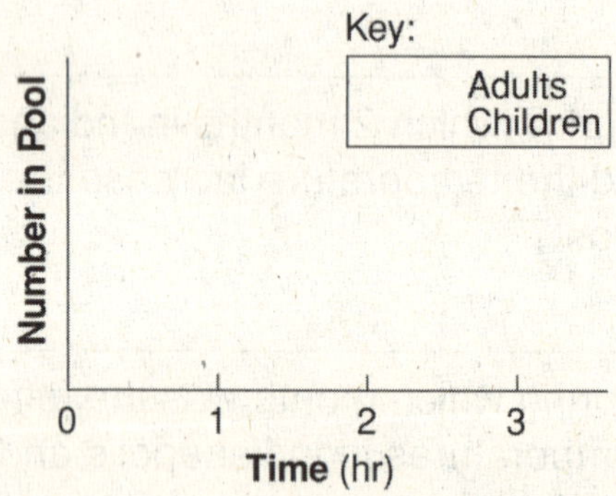

6. The Ice Cream Station is open only during the summer months. Cool Guy's Ice Cream is open all year long. Show the change in ice cream sales over the course of a year for each business.

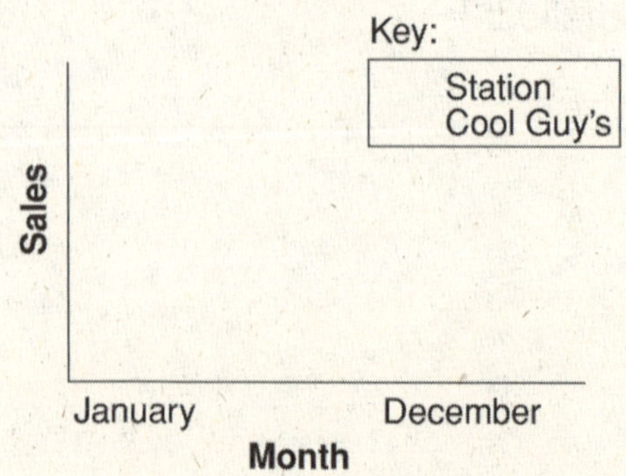

Name ______________________ Date ______________ Class ______________

LESSON 7-6

Problem Solving

Line Graphs

Write the correct answer.

The line graph shows the number of households with cable television from 1996 to 2002.

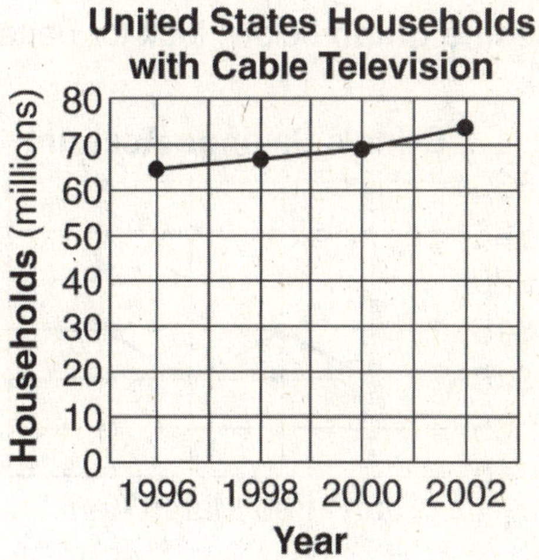

1. About how many households had cable TV in 1998?

____

2. About how many more households had cable TV in 2002 than in 1996?

____

3. During which two-year period did the number of households with cable TV grow the most?

____

4. Use the graph to estimate the number of households with cable TV in 2001.

____

Choose the letter for the best answer.

The double-line graph shows the number of tornadoes in the United States during part of 2002 and 2003.

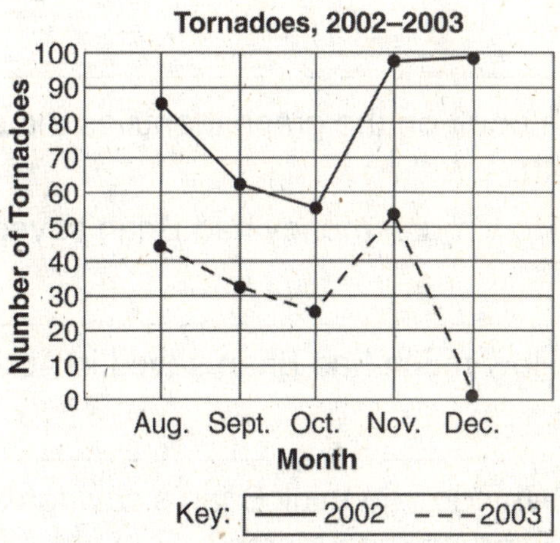

5. About how many more tornadoes were there in August 2002 than in August 2003?

   A about 50　　C about 30

   B about 40　　D about 20

6. In which time period did the number of tornadoes increase both years?

   F Sept. to Oct.　　H Nov. to Dec.

   G Oct. to Nov.　　J none

7. Which time period showed the greatest decrease in the number of tornadoes?

   A Aug. to Sept. 2002

   B Aug. to Sept. 2003

   C Oct. to Nov. 2002

   D Nov. to Dec. 2003

8. During which month was there the greatest difference between the number of tornadoes in 2002 and in 2003?

   F Aug.　　H Nov.

   G Sept.　　J Dec.

Name ______________________ Date ______________ Class ______________

LESSON 7-6

Reading Strategies

Organization Patterns

A **line graph** visually shows data that changes over time. Your height changes from year to year. Sales at a store change from month to month.

In the **line graph** below, look for patterns of change in a savings plan.

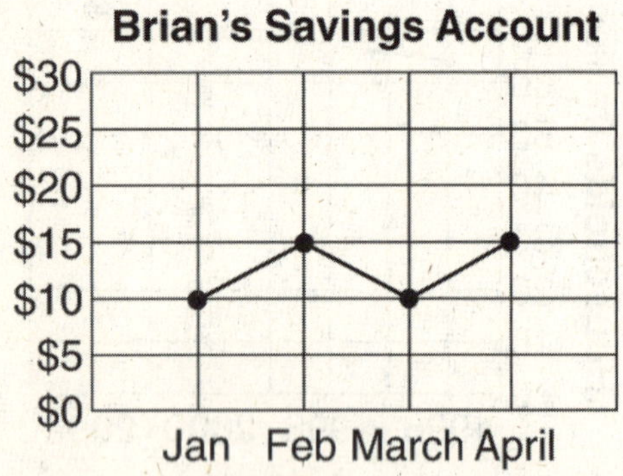

Answer the following questions about the line graph.

1. What information is located along the left side of the graph?

____

2. In what intervals are the numbers shown on the left side of the graph?

____

3. What information is shown along the bottom of the graph?

____

Each point on the graph identifies the amount of money saved each month.

4. How much money had Brian saved in February?

____

5. How much had Brian saved in April?

____

6. Why do you think Brian's savings account dropped between February and March?

____

Name ______________________ Date ______________ Class ______________

LESSON 7-6

Puzzles, Twisters & Teasers

Rhyme Time TV!

Make a line graph of the data in the table. Think about the data and the way it is presented. Then make a rhyme to answer the riddle by filling in the letter associated with each ordered pair.

What do you call a table that shows information about cable TV?

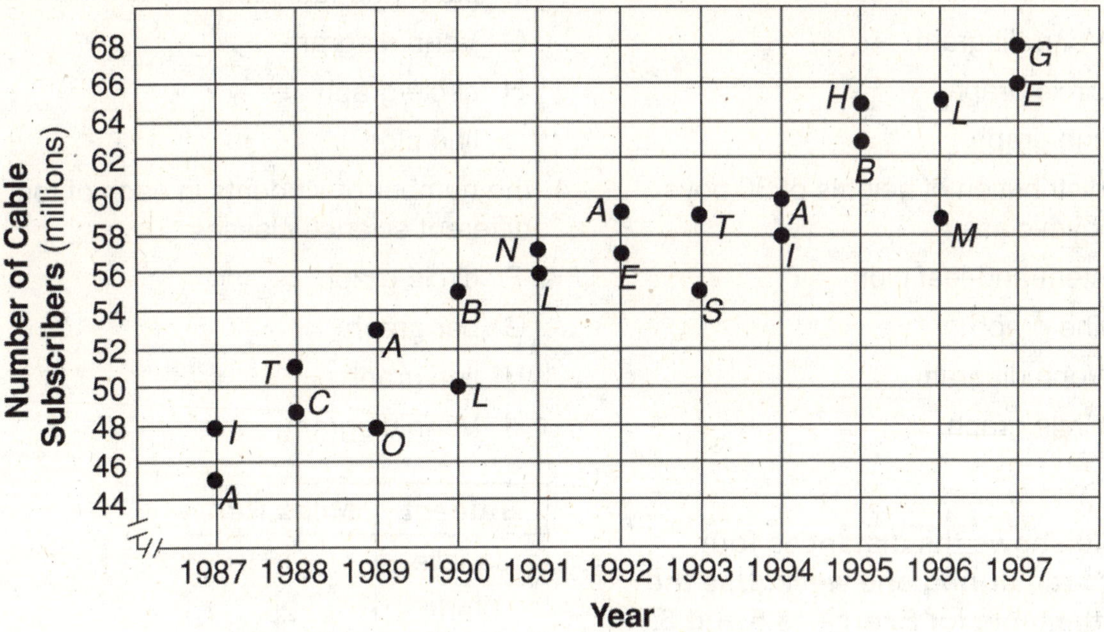

Year	Number of Cable Subscribers (millions)	Letter to Solve Riddle
1987	45	
1988	49	
1989	53	
1990	55	
1991	56	
1992	57	
1993	59	
1994	60	
1995	63	
1996	65	
1997	66	

Name ______________________ Date ______________ Class ______________

LESSON 7-7

Practice A

Choosing an Appropriate Display

Choose the letter of the graph that would best represent each type of data.

1. the number of visitors to a national park in each of the past five years
   - A stem-and-leaf plot
   - B Venn diagram
   - C circle graph
   - D line graph

2. the part of a whole group that said yes, no, or no opinion in a survey
   - F stem-and-leaf plot
   - G Venn diagram
   - H circle graph
   - J line plot

3. the distribution of heights of 30 boys in a gym class
   - A stem-and-leaf plot
   - B line graph
   - C Venn diagram
   - D circle graph

4. the number of students in each of four different science classes
   - F circle graph
   - G bar graph
   - H line graph
   - J Venn diagram

The table shows the distances four students ran during one week. Use the data in the table for Exercises 5 and 6.

Student	Miles Run
Janelle	15
Marty	25
Rosa	30
Sidney	10

5. Does the stem-and leaf plot do a good job of representing the data from the table? Explain your answer.

Distances Run

Stems	Leaves
1	0 5
2	5
3	0

Key: 1|2 means 12 miles

____

____

____

6. Does the bar graph do a good job of representing the data? Explain.

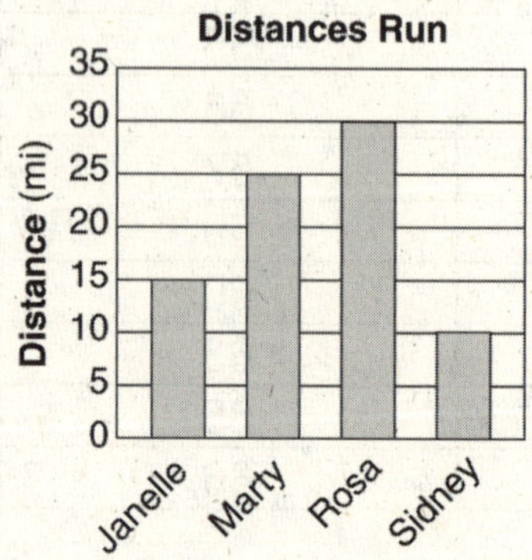

____

____

____

Name ______________________ Date ______________ Class ______________

LESSON 7-7

Practice B

Choosing an Appropriate Display

Choose the type of graph that would best represent this data.

1. the number of points scored by five different basketball players in a game

____

2. the distribution of test scores in a math class

____

3. the students who are in the chess club, the debating club, and the computer club, and the students who are in more than one of those clubs

____

4. the percent of total income a family uses for rent, food, clothing, entertainment, savings and other expenditures

____

The table shows the earnings of The Sandman Company during each of five years. Explain why each display does or does not appropriately represent the data.

Year	Earnings (millions of dollars)
2001	1
2002	3
2003	8
2004	12
2005	16

5.

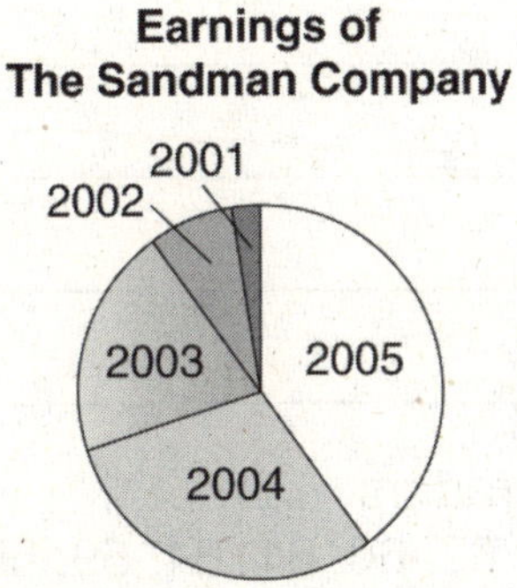

____

6.

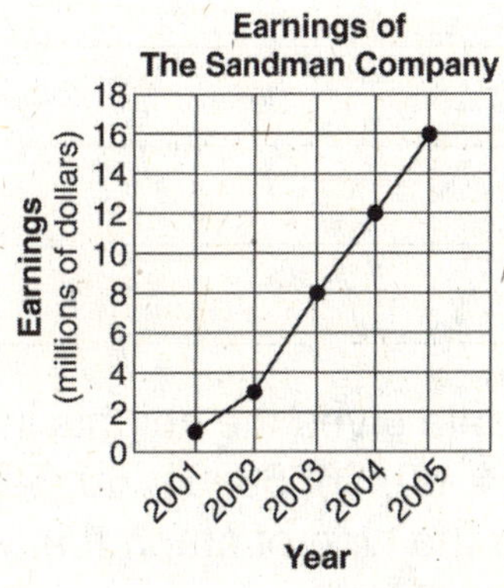

____

Name ______________________ Date ______________ Class ______________

LESSON 7-7

Practice C

Choosing an Appropriate Display

Choose the type of graph that would best represent this data.

1. the distribution of scores of a professional golfer for 20 rounds of golf

____

2. the change in weight over time when the diet of a dog is changed.

____

The table shows the responses of 100 students to a survey. Explain why each display does or does not appropriately represent the data.

Favorite Sport to Play	Number of Students
Baseball	15
Basketball	30
Football	10
Soccer	40
Tennis	5

3.

____

____

____

____

____

4.

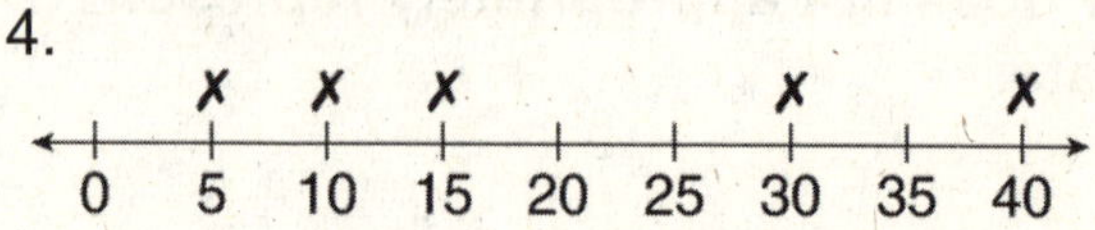

____

____

____

____

____

5. The data gives the heights in inches of 20 seventh-graders.
58, 64, 60, 56, 66, 61, 60, 58, 70, 62, 60, 58, 56, 71, 62, 60, 61, 64, 59, 56
Make the type of graph that would best represent the data.

Name ______________________ Date ______________ Class ______________

LESSON 7-7

Review for Mastery

Choosing an Appropriate Display

There are different ways to show data. To choose a data display, think about the information that each kind of graph shows.

Bar graph: Compares amounts.

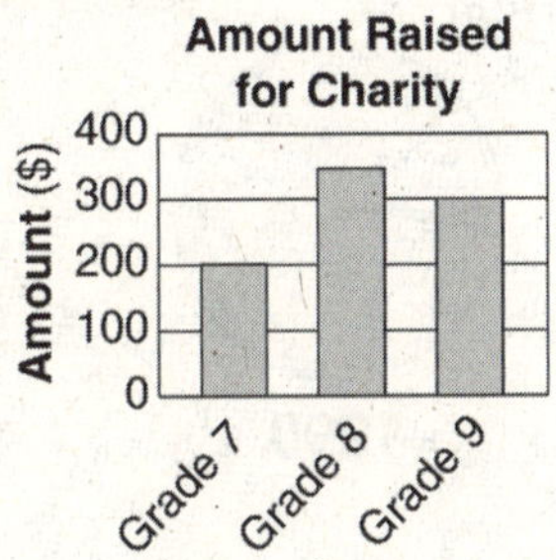

Circle graph: Shows data as parts of a whole.

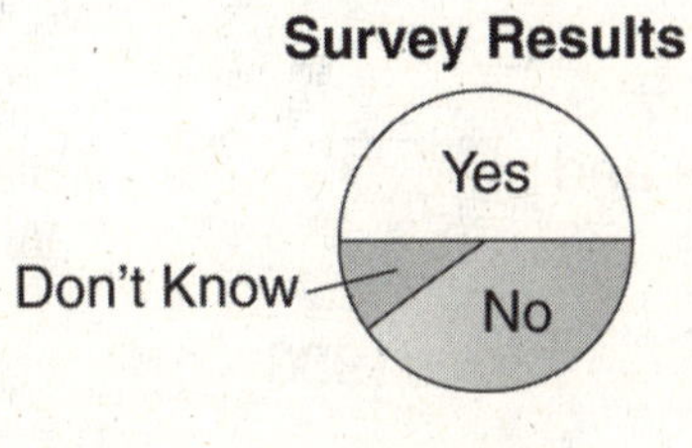

Stem-and-leaf plot: Shows each value and how the values are distributed.

Test Scores

Stems	Leaves
6	8
7	5 6 6 8 9 9
8	2 2 4 4 5 8 8
9	1 7

Key: 6|8 means 68

Line plot: Shows each value and how the values are distributed.

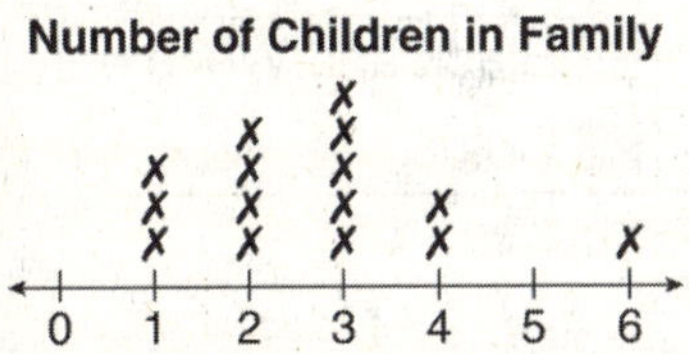

Line graph: Shows how amounts change over time.

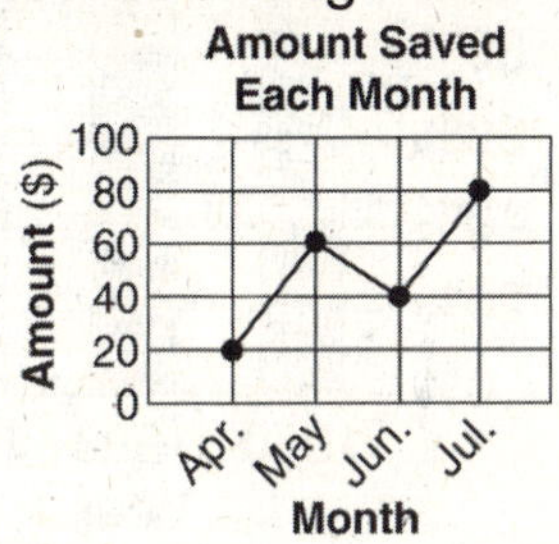

Venn diagram: Shows members of sets and how sets overlap.

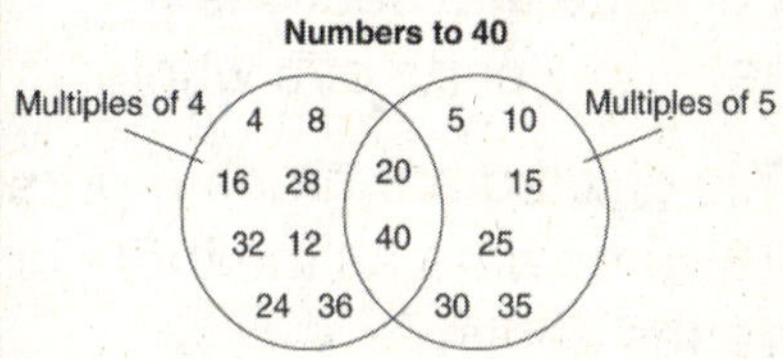

Circle the type of graph which would best show the data. Explain the reason for your choice.

1. the way the vote is split among three candidates in a school election: circle graph or stem-and-leaf plot?

____

____

____

2. the points scored by a team during each game in a 16-game season: stem-and-leaf plot or circle graph?

____

____

____

3. the average income for a family of four in 1970, 1980, 1990, and 2000: line plot or line graph?

____

____

____

4. the calories per ounce in four different cuts of meat: line plot or bar graph?

____

____

____

Name ______________________ Date ______________ Class ______________

LESSON 7-7

Challenge

Divided Bar Graphs

Suppose you wanted to show how parts of whole change over time. One way to do this would be to draw a series of circle graphs.

These circle graphs show the percent of TV households that have color TVs or only black and white TVs.

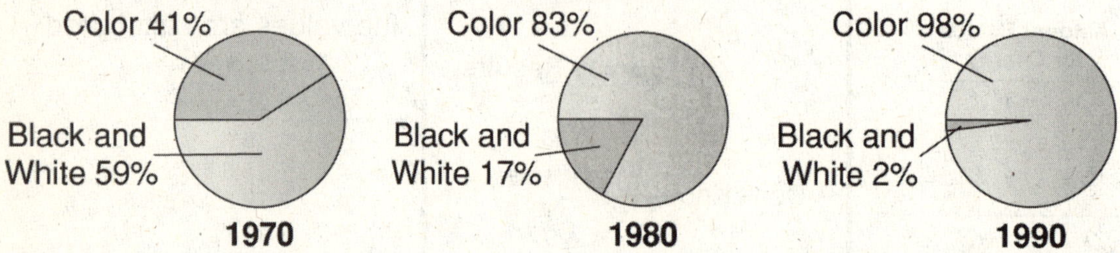

Another display you could use is a divided bar graph. In a divided bar graph, each bar is divided into sections that show parts of a whole.

This divided bar graph represents the same data as the three circle graphs above.

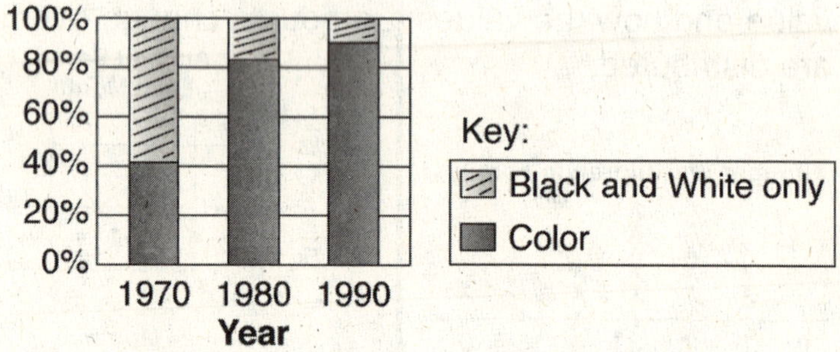

The table shows how many hours of television students at one high school watched. Use the data from the table to complete the divided bar graph.

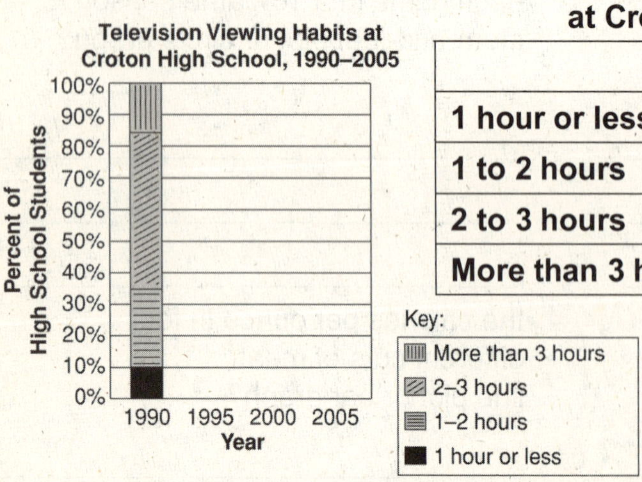

Television Viewing Habits at Croton High School, 1990–2005

	1990	1995	2000	2005
1 hour or less	40	44	70	120
1 to 2 hours	100	124	175	180
2 to 3 hours	200	176	80	50
More than 3 hours	60	56	50	30

Name ______________________ Date ______________ Class ______________

LESSON 7-7

Problem Solving

Choosing an Appropriate Display

Write the correct answer.

1. You take a survey of your class to find the number of years each student has lived in your town. You want to show how the data is distributed. What kind of graph would you use to display your data? Explain your choice.

____

____

____

2. Wendy budgets $120 for the week. $30 is for transportation, $50 is for food, $25 is for entertainment, and $15 is for other expenses. What kind of graph would best represent the data? Justify your response.

____

____

____

3. The New York Mets' worst season was their first season, in 1962. In that year they won 40 games. Their best season was in 1986, when they won 108 games. You want to show the number of games won each year in a way that makes it easy to see the distribution of the data. Why might you use a stem-and-leaf plot rather than a line plot?

____

____

____

4. During one school year, Matt reads 16 books, Tama reads 22 books, Rhonda reads 14 books, and Francisco reads 20 books. Would a line graph be an appropriate way to display this data? Explain your answer.

____

____

____

Choose the letter for the best answer.

5. Which type of graph would be most appropriate to show the distribution of daily high temperatures for a month?
   - A circle graph
   - B Venn diagram
   - C stem-and-leaf plot
   - D line graph

6. Which type of graph would be most appropriate to compare the price of the same TV at five discount stores?
   - F stem-and-leaf plot
   - G bar graph
   - H circle graph
   - J line plot

Name ______________________ Date ______________ Class ______________

LESSON 7-7

Reading Strategies

Use a Flowchart

You can use a flowchart to help you choose an appropriate display for a set of data.

Start Here

Do you need to show amounts that change over time?	**Yes** →	Use a line graph.
↓ **No**		
Do you need to compare amounts?	**Yes** →	Use a bar graph.
↓ **No**		
Do you need to show parts of a whole?	**Yes** →	Use a circle graph.
↓ **No**		
Do you need to show two or more sets of data and how they overlap?	**Yes** →	Use a Venn diagram.
↓ **No**		
Do you need to see how often each piece of data occurs?	**Yes** →	Use a line plot or a stem-and-leaf plot.

Use the flowchart and the situation below for Exercises 1 and 2.

For a TV show, 55% of the viewers are under 18, 35% are between 18 and 34, and 10% are over 34. You want to show how big a part of the whole audience each age group is.

1. For which question in the flowchart is the answer yes?

____

2. Which kind of graph will you use? ______________

Use the flowchart and the situation below for Exercises 3 and 4.

A school has 34 different classes with sizes between 18 and 28 students. You want to show each of the different class sizes.

3. For which question in the flowchart is the answer yes?

____

4. Which kind of graph will you use? ______________________

Name ______________________ Date ______________ Class ______________

LESSON 7-7

Puzzles, Twisters & Teasers

Seeing Is Believing!

Choose the type of graph that would be best for each situation. Circle the letter above your answer.

1. comparing the numbers of raffle tickets sold by four students

   L line plot — **O** bar graph

2. displaying every winning speed in the history of the Indianapolis 500 car race

   Y stem-and-leaf plot — **D** circle graph

3. showing the factors and common factors of 24, 30, and 35

   A circle graph — **G** Venn diagram

4. showing the sales of a monthly magazine in its first six months

   R Venn diagram — **E** line graph

5. comparing the parts of a monthly budget

   D circle graph — **T** stem-and-leaf plot

6. displaying quiz scores for 20 quizzes on which the scores can be from 1–10

   S line plot — **P** line graph

Use the circled letters above the problem numbers to solve the riddle.

What do you need to spot an iceberg from 10 miles away?

___ ___ ___ ___   ___ ___ ___ ___
3. 1. 1. 5.   4. 2. 4. 6.

Name ______________________ Date ______________ Class ______________

LESSON 7-8

Practice A

Populations and Samples

Choose the letter of the sampling method that will better represent the whole population.

1. Clinton School Cafeteria: Student Satisfaction

   a. Mark surveys 40 students who are in his classes. 72% are satisfied with the food in the cafeteria.

   b. Tammy surveys 65 students by randomly choosing names from a list of all students in the school. 85% are satisfied with the food in the cafeteria.

2. Predicted Winner in an Election for Mayor

   a. Harris telephones 100 randomly chosen voters. 54% plan to vote for Mayor Green.

   b. Julia asks 70 people whom she knows. 45% plan to vote for Mayor Green.

For Problems 3–5, tell whether each sample may be biased. Explain your answer.

3. A town official surveys 50 people in a library to decide if town residents want the library expanded.

   ____

   ____

4. A cable TV company randomly calls 200 customers and asks them if they are satisfied with their service.

   ____

5. George surveys 15 students on the soccer team to learn whether middle school students want more money spent on school sports.

   ____

   ____

   ____

6. A factory produces 12,000 computers per week. The manager of a factory claims that fewer than 50 defective computers are produced each week. In a random sample of 500 computers, 2 were defective. Tell if the manager's claim is likely to be true.

   ____

Name ______________________ Date ____________ Class ____________

LESSON 7-8

Practice B

Populations and Samples

1. Determine which sampling method will better represent the entire population. Justify your answer.

Reading Habits of High School Students	
Sampling Method	**Results of Survey**
Dinah surveys 48 students who she knows.	91% have read a novel in the past month.
Suki gives survey forms to 100 students who were randomly chosen from a school attendance list.	59% have read a novel in the past month.

____

____

____

For Problems 2 and 3, determine whether each sample may be biased. Explain.

2. An on-line bookseller randomly chooses 200 book buyers from its database and then surveys those book buyers to find out if they were satisfied with the time it took to deliver their orders.

____

____

3. Milena surveys 80 high school students who are leaving a jazz concert to determine the favorite type of music among high school students.

____

____

4. Zack chooses a random sample of 50 out of 400 students. He finds that 7 of them have traveled to a foreign country. Zack claims that over 50 of the 400 students have traveled to a foreign country. Do you agree? Explain your answer.

____

5. A mint produces 150,000 souvenir coins each year. In a random sample of 400 coins, 3 have a misprint. Predict the number of coins that will have misprints in a year.

____

Name ______________________ Date ______________ Class ______________

LESSON 7-8

Practice C

Populations and Samples

1. Determine which sampling method will better represent the entire population. Justify your answer.

The Midland Company: Employee Satisfaction	
Sampling Method	**Results of Survey**
Wanda interviews 80 employees who were randomly selected from the company payroll list.	69% feel adequately challenged by their jobs
Bernard interviews the last 30 employees to be hired.	90% feel adequately challenged by their jobs.

____

____

____

For Problems 2 and 3, determine whether each sample may be biased. Explain.

2. A landlord e-mails 60 of his 1,250 tenants and surveys them to determine whether they would like to use the Internet to pay rent.

____

____

3. An insurance company surveys 350 of its customers by randomly choosing names from its customer database and then telephoning the customers.

____

Explain whether you would survey the entire population or use a sample.

4. You want to know how many hours members of a sports team train each week during the off-season.

____

5. You want to know the average income of people who eat at vegetarian restaurants across the country.

____

Name ______________________ Date ______________ Class ______________

LESSON 7-8

Review for Mastery

Populations and Samples

Survey topic: number of books read by seventh-graders in Richmond

A **population** is the whole group that is being studied.	*Population*: all seventh-graders in Richmond
A **sample** is a part of the population.	*Sample*: all seventh graders at Jefferson Middle School
A **random sample** is a sample in which each member of the population has a random chance of being chosen. A random sample is a better representation of a population than a non-random sample.	*Random sample*: Have a computer select every tenth name from an alphabetical list of each seventh-grader in Richmond.
A **biased sample** is a sample that does not truly represent a population.	*Biased sample*: all of the seventh graders in Richmond who are enrolled in honors English classes.

Tell if each sample is biased. Explain your answer.

1. An airline surveys passengers from a flight that is on time to determine if passengers on all flights are satisfied.

____

____

2. A newspaper randomly chooses 100 names from its subscriber database and then surveys those subscribers to find if they read the restaurant reviews.

____

3. The manager of a bookstore sends a survey to 150 customers who were randomly selected from a customer list.

____

4. A team of researchers surveys 200 people at a multiplex movie theater to find out how much money state residents spend on entertainment.

____

____

____

Name ______________________ Date ______________ Class ______________

LESSON 7-8

Challenge

A Sample of Samples

Consider the three sampling methods described below.

Systematic sample: You randomly select an individual and then follow a pattern to select others in the sample. For example, choose a name from the first 50 names in a telephone book. Then choose every fiftieth name after that.

Stratified sample: A population is divided into subgroups, called strata, that contain similar individuals. For example, interview all the boys and girls at a movie. Boys and girls are two strata.

Cluster sample: A population is divided into sections and then a few of the sections are selected. All the members of those sections are chosen. For example, suppose a city has 50 voting districts. Five of those districts are selected, and every person in each district selected is polled.

Identify the type of sampling used as systematic, stratified, or cluster.

1. A teacher selects every third student in the class. ______________
2. A teacher surveys all students from each of 5 randomly selected classes. ______________
3. A principal selects 3 girls and 3 boys from each of 10 classes. ______________
4. A teacher selects 10 students under 12 years old and 10 students over 12 years old. ______________
5. A reporter interviews all students in each of 4 randomly selected schools. ______________
6. A reporter interviews 20 men and 20 women. ______________
7. Workers on an assembly line check every tenth tire in a tire manufacturing plant. ______________
8. An advertising consultant surveys all members in each of 3 randomly selected fitness clubs from all the fitness clubs in the city. ______________
9. A department store manager selects a customer from the first 20 on a customer list. Then she selects every twentieth customer after that. ______________
10. Members of the school board survey 25 elementary school students, 25 middle school students, and 25 high school students. ______________

Name ______________________ Date ______________ Class ______________

LESSON 7-8

Problem Solving

Populations and Samples

Write the correct answer.

1. Max wants to find out the exercise habits of local children. He plans to survey every third child he sees coming out of a sporting goods store. Max says his sample is not biased. Do you agree? Explain your answer.

____

____

____

2. Ms. Constantine is choosing among three field trips for her two classes. She wants to determine which trip her students prefer. Should she survey the entire population or use a sample? Explain.

____

____

____

3. A researcher catches 60 fish from different locations in a lake. He then tags the fish and puts them back in the lake. Two weeks later, the researcher catches 40 fish from the same locations. 8 of these 40 fish are tagged. Predict the number of fish in the lake.

____

4. A high school has 1,800 students. A random sample of 80 shows that 24 have cell phones. Predict the number of students in the high school who have cell phones.

____

Choose the letter for the best answer.

5. The school board wants to study computer literacy among teachers. Which would represent a random sample of teachers?
   - A all high school math teachers
   - B teachers from the middle school whose name begins with N
   - C all male teachers
   - D every eighth teacher on an alphabetical list

6. In a random sample, 3 of 400 computer chips are found to be defective. Based on the sample, about how many chips out of 100,000 would you expect to be defective?
   - F 750
   - G 3,000
   - H 4,000
   - J cannot be determined

Name ______________________ Date ______________ Class ______________

LESSON 7-8

Reading Strategies

Compare and Contrast

To get information about issues, a survey is conducted. Surveys can be done in two different ways.

- **Population** The entire group is surveyed.
- **Sample** Part of the entire group is surveyed.

1. Compare the difference between collecting information from the population and collecting a sample.

____

There are two different types of samples.

- **Unbiased sample** The sample represents the population.
- **Biased sample** The sample does not represent the population.

2. What is the difference between an unbiased sample and a biased sample?

____

____

Mrs. Jones wants to know which sport 7th graders in the district like best. There are 7th graders in 6 different schools in the district. She can collect data in one of the following ways:

Population—Ask every 7th grade student at all 6 schools.
Unbiased sample—Ask every other 7th grader at 3 of the schools.
Biased sample—Ask 7th grade boys at 3 of the schools.

Write "unbiased sample" **or** "biased sample" **to describe each survey.**

3. A survey conducted at an ice cream store asked only mothers their favorite ice cream flavor.

____

4. A reporter asked every tenth person coming out of a theater how well they liked the movie.

____

5. A survey asked only girls to identify their favorite item on the school cafeteria menu.

____

Name ______________________ Date ______________ Class ______________

LESSON 7-8

Puzzles, Twisters & Teasers

Answer This!

Choose the best description of each sample. Circle the letter above your answer.

1. Tyler surveys 60 people at a high school football game to find out which high school sports team is most popular.

O	**A**
biased	not biased

2. A computer manufacturer randomly calls 150 people who called its help line and surveys them to determine if they were satisfied with the help they received.

M	**N**
biased	not biased

3. Margie surveys 12 of her friends to determine what percentage of students participate in after-school clubs.

F	**D**
random sample	convenience sample

4. Tamira surveys 50 riders on one commuter train to determine whether riders on all trains in the system think that the on-time performance is adequate.

K	**A**
biased	not biased

5. Sam calls every sixth student on a high school attendance list and surveys the students to determine their favorite subject.

M	**R**
random sample	convenience sample

Write the circled letters above the problem numbers to solve the riddle.

What is the difference between dinosaurs and dragons?

Dinosaurs ____ ____ ____ ____ T S ____ ____ ____ E

3. 1. 2. 1. 5. 1. 4.

Name ______________________ Date ______________ Class ______________

LESSON 7-9

Practice A

Scatter Plots

The table shows how much energy was produced from wind power in the United States from 2000 to 2004.

Year	2000	2001	2002	2003	2004
Wind Power (trillions of BTUs)	57	70	105	115	143

1. Make a scatter plot of the data.

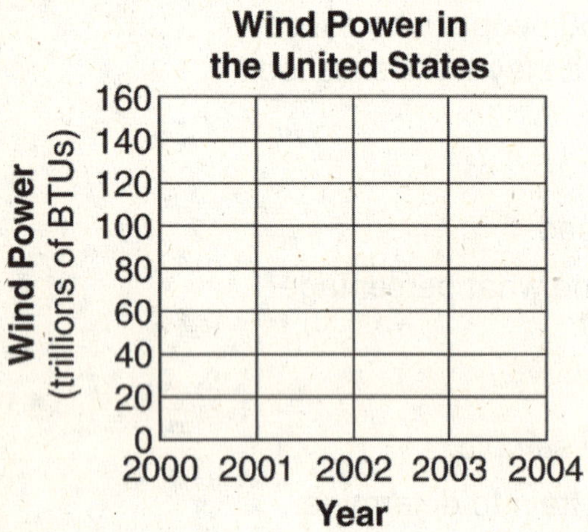

2. What does the scatter plot show about how much wind power was produced in the United States from 2001 to 2004?

____

____

3. What kind of correlation does the scatter plot show—positive, negative, or no correlation?

____

Write *positive, negative,* or *no correlation* to describe each relationship.

4.

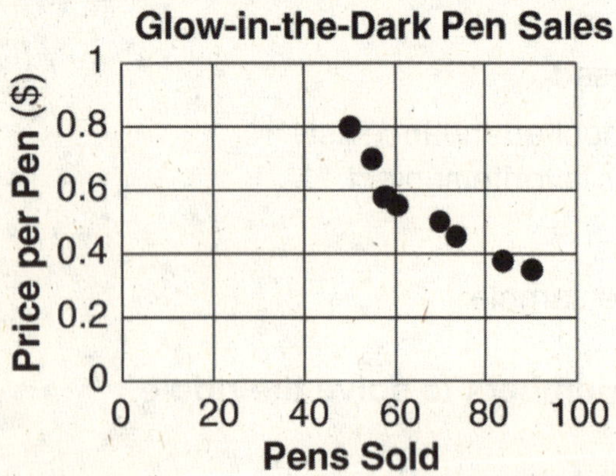

____

5.

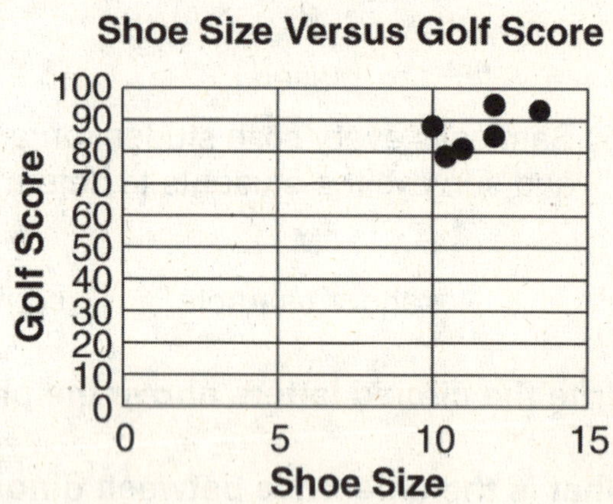

____

6. outside temperature and the chance of your getting frostbite

____

7. number of hours spent studying for a test and the test scores

____

8. number of basketballs going into the basket and distance in feet from the basket when shooting

____

9. height and age of adults over 25

____

Name ______________________ Date ______________ Class ______________

LESSON 7-9

Practice B

Scatter Plots

The table shows boys' average heights in inches from ages 6 through 13. Use the table for Exercises 1–3.

Age	6	7	8	9	10	11	12	13
Height (in.)	$46\frac{3}{4}$	49	51	$53\frac{1}{4}$	$55\frac{1}{4}$	$57\frac{1}{4}$	59	61

1. Make a scatter plot of the data.

2. Describe the relationship between the data sets.

____

____

3. What kind of correlation does the plot show?

____

Write *positive, negative,* or *no correlation* to describe each relationship.

4.

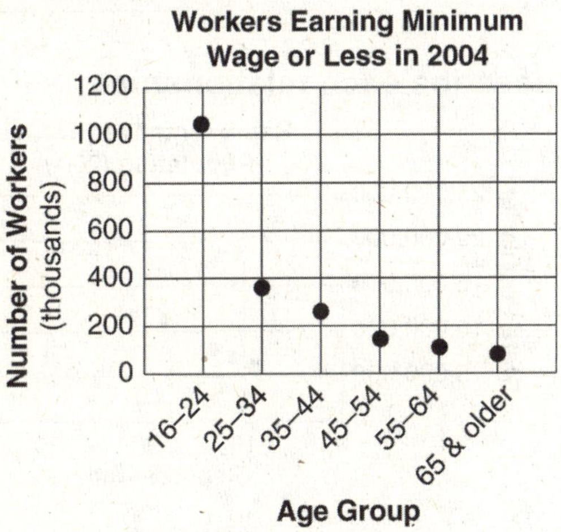

____

5.

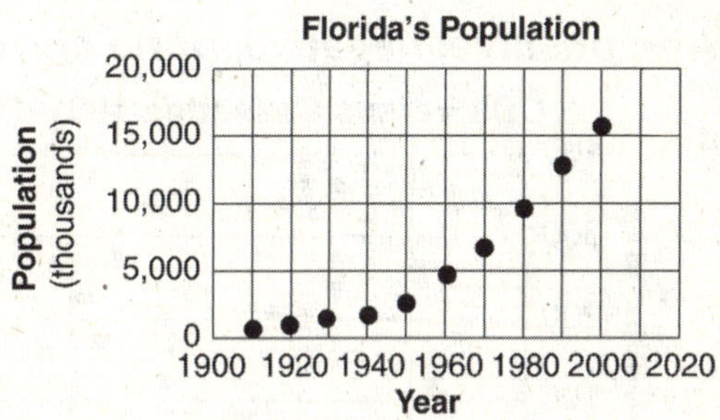

____

6. student test scores and the number of students who walk to school

____

7. the grade levels of students and their ages in months

____

8. the year a state entered the union and the number of years as a state

____

9. ages of students and their grades on tests

____

Name ______________________ Date ______________ Class ______________

LESSON 7-9

Practice C

Scatter Plots

The table shows the percentage of homes with a PC based on household income in 2001. Use the table for Exercises 1–3.

1. Make a scatter plot of the data.

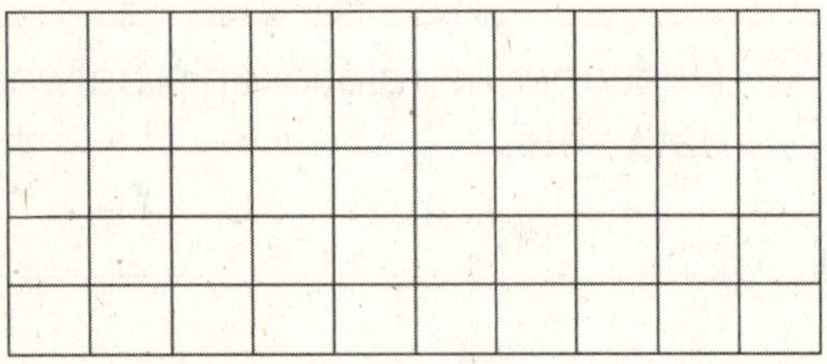

Annual Income	Homes with PCs
Less than $5,000	26%
$5,000–$9,999	19%
$10,000–$14,999	26%
$15,000–$19,999	32%
$20,000–$24,999	40%
$25,000–$34,999	50%
$35,000–$49,999	64%
$50,000–$74,999	78%
$75,000 and up	89%

2. Describe the relationship between the data sets.

____

____

3. What kind of a correlation does the plot show? ______________

Write *positive, negative,* or *no correlation* to describe each relationship.

4.

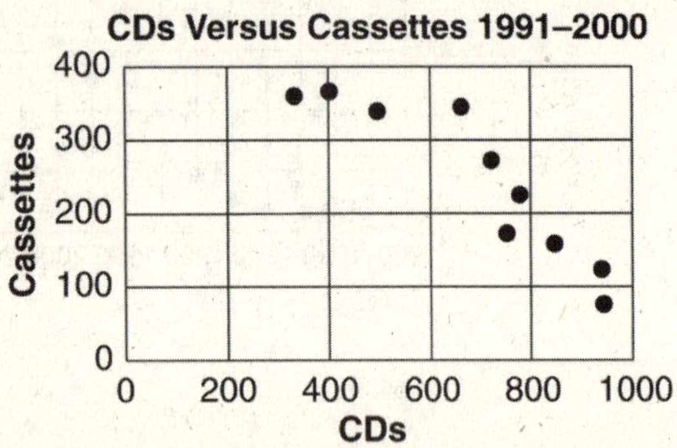

____

5.

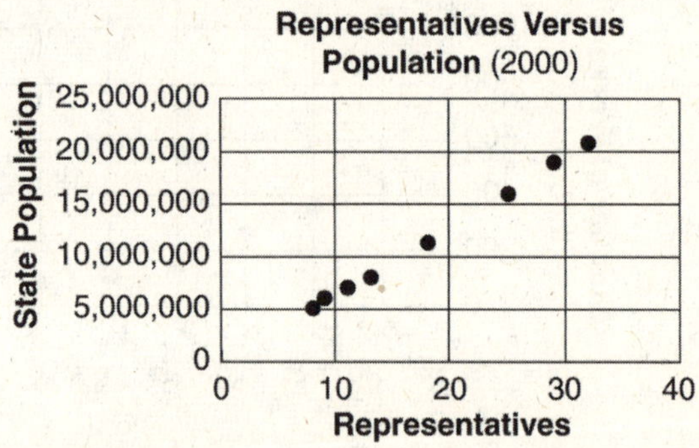

____

6. the number of senators and the state population

____

7. the number of hours spent exercising and the number of calories burned

____

Name ________________________ Date ______________ Class ______________

LESSON 7-9

Review for Mastery

Scatter Plots

A **scatter plot** can be used to explore how two sets of data are related.

A scatter plot can show three different relationships between data sets.

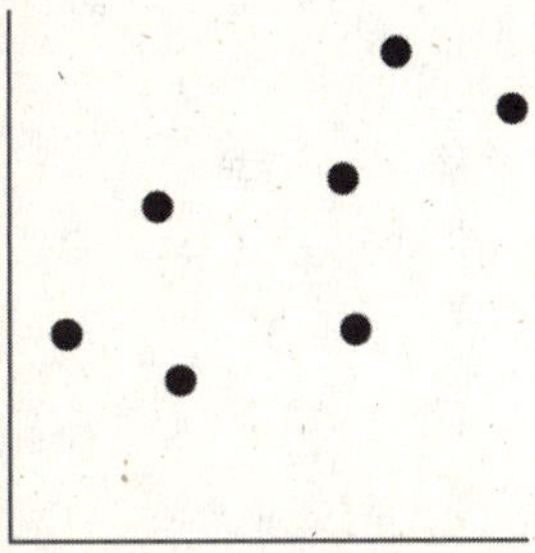

Positive Correlation
y-axis values increase as *x*-axis values increase.

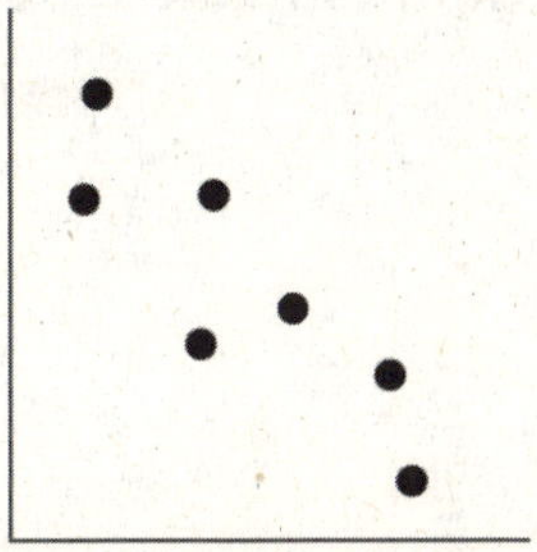

Negative Correlation
y-axis values decrease as *x*-axis values increase.

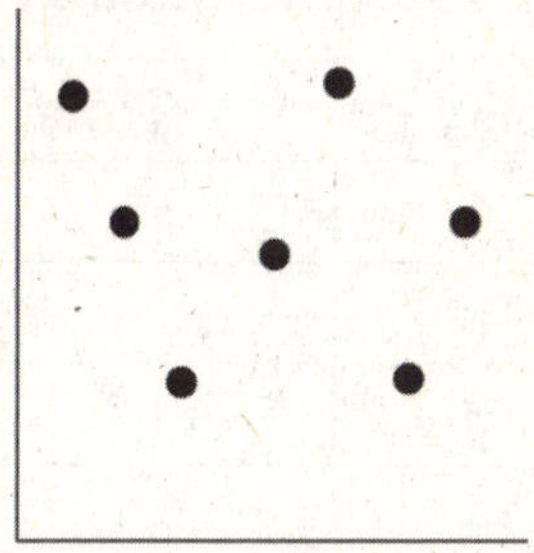

No Correlation

Look at each scatter plot. What kind of correlation is shown?

1. **Feeding a Family**

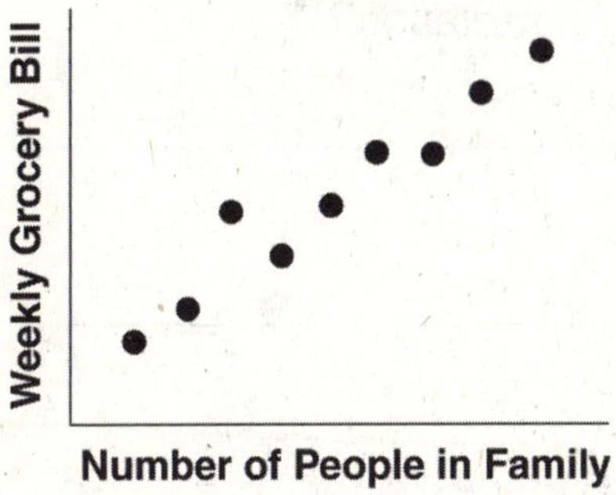

____

2. **Science and Your Feet**

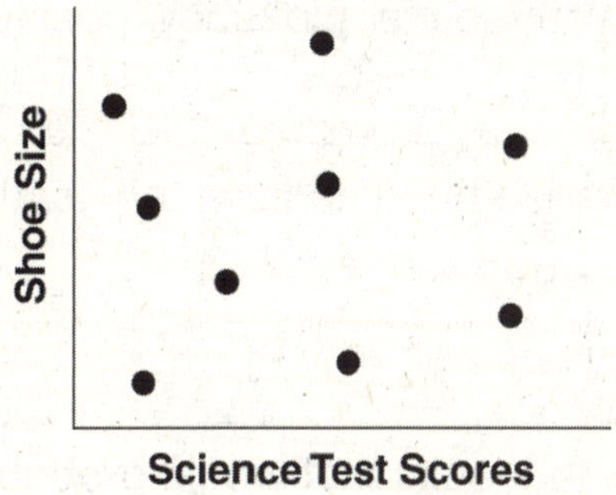

____

What type of correlation might you expect between:

3. the number of pages in a book and the number of copies the book sells?

____

4. the number of computers per school and the number of students per computer?

____

5. the number of students in a city and the number of students who play team sports?

____

6. the temperature outside and the amount of heat used inside?

____

7. a person's height and his or her birthday?

____

8. the speed of a car and the time required to travel a certain distance?

____

Name ______________________ Date ______________ Class ______________

LESSON 7-9

Challenge

Time On Task

Use a stopwatch to find out how long it takes to write your first and last names backwards. Do it 10 times, filling in the table with the time of each try. Use the data to make a scatter plot.

Try	Time
1	
2	
3	
4	
5	
6	
7	
8	
9	
10	

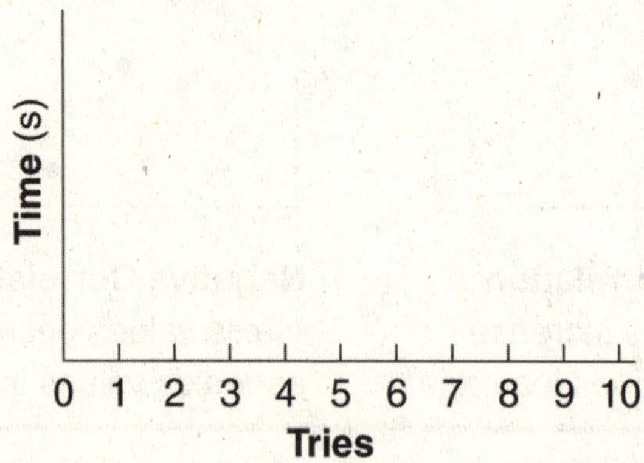

1. Does your scatter plot show positive, negative, or no correlation?

____

2. Summarize the data from the scatter plot.

____

____

3. Check the tasks that would probably have a scatter plot that is similar to the one you drew above after doing each task 10 times.

Making tacos in a fast food restaurant		Writing multiplication facts from 1 to 12	
Filling boxes on an assembly line		Eating dinner	
Driving from home to school		Talking to a friend on the phone	

4. Based on the data and scatter plot, what conclusion can you make about learning to do a job?

____

____

Name ______________________ Date ______________ Class ______________

LESSON 7-9

Problem Solving

Scatter Plots

Write the correct answer.

This scatter plot compares the mean annual income of Americans with the number of years spent in school.

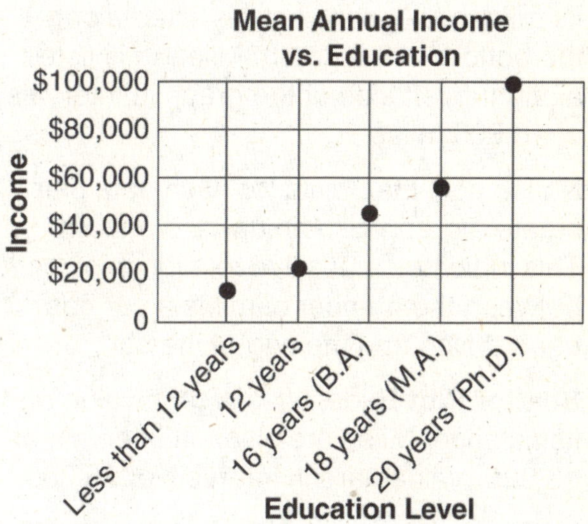

1. Which level of education has a mean annual income between $40,000 and $50,000?

____

2. Estimate the range of income data on this scatter plot.

____

3. Which level of education has the lowest income?

____

4. Does the scatter plot show a positive correlation, negative correlation, or no correlation between education and income?

____

Choose the letter for the best answer.

5. What kind of correlation would you expect to find between a city's annual snowfall amount and the size of its population?

   A positive correlation

   B negative correlation

   C no correlation

   D impossible to say

6. What kind of correlation would you expect to find between a movie's length and the number of times it can be shown in a day?

   F positive correlation

   G negative correlation

   H no correlation

   J impossible to say

7. What kind of correlation would you expect to find between an animal's mass and the number of calories it consumes in a day?

   A positive correlation

   B negative correlation

   C no correlation

   D impossible to say

8. What kind of correlation would you expect to find between a person's height and his or her income?

   F positive correlation

   G negative correlations

   H no correlation

   J impossible to say

Name ______________________ Date ______________ Class ______________

LESSON 7-9

Reading Strategies

Drawing Conclusions

You can place individual data points on a graph to see if a pattern occurs. This type of graph is called a **scatter plot**.

Scatter Plot A identifies a car's age and its mileage. Age in years is listed along the bottom of the graph. Mileage is listed along the left side of the graph in intervals of 10,000 miles.

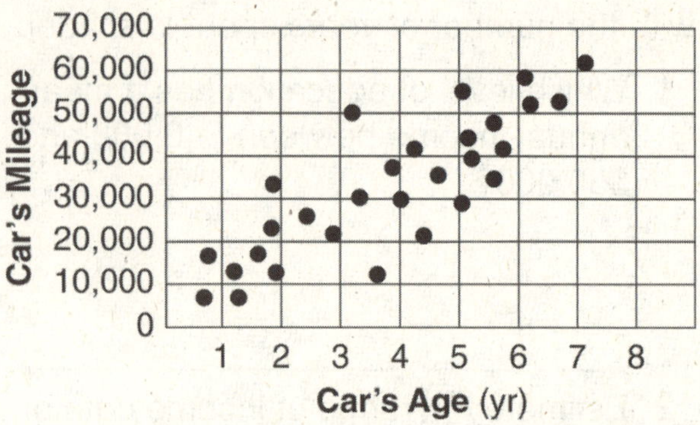

Notice how the points for each data pair create a graph with an upward pattern. This indicates a positive relationship between the number of miles a car has been driven and the age of the car.

Scatter Plot B identifies a car's value and its mileage. Miles are given in intervals of 10,000. Values are in intervals of $5,000.

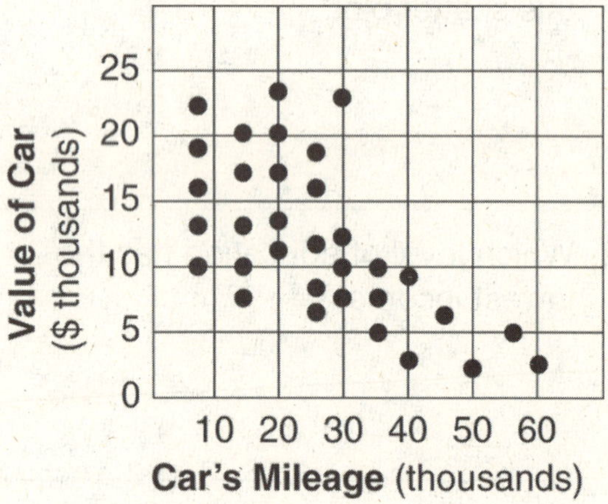

Notice how the points for each data pair create a downward pattern. This shows that the greater the mileage, the less a car is worth.

Scatter Plot C identifies a car's age and the number of passengers in the car.

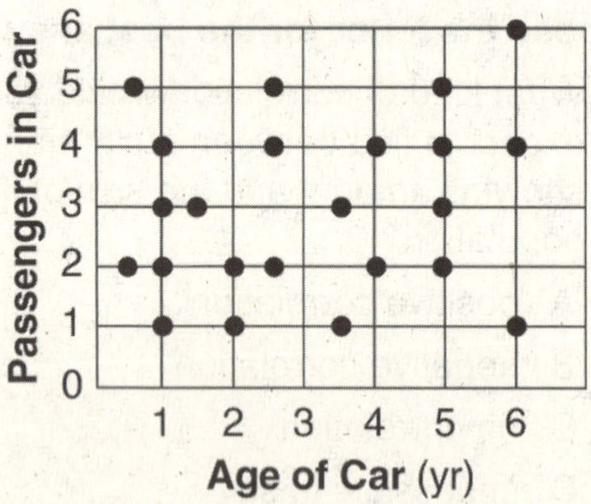

Notice how the points are scattered on the graph. This shows that there is no relationship between these two sets of data.

Read each statement. Identify the scatter plot that helps you determine the answer, and write "true" or "false."

1. The older the car, the fewer the miles. ______________________
2. The more miles on a car, the less its value. ______________________
3. The newer the car, the fewer the miles. ______________________
4. The older the car, the more passengers it can carry. ______________________

Name ______________________ Date ______________ Class ______________

LESSON 7-9

Puzzles, Twisters & Teasers

Scatter Brained!

Does the size of the brain determine the size of the body? Chart the data from the table below as a scatter plot. Then answer the questions. Use the underlined letters in the correct answers to solve the riddle.

Animal	**Brain Weight** (oz)	**Body Weight** (lb)
Human	53	140
Bottlenose Dolphin	56	340
White Shark	1.2	950
Alligator	8.5	550
Cow	17	1,000
Sheep	5	150

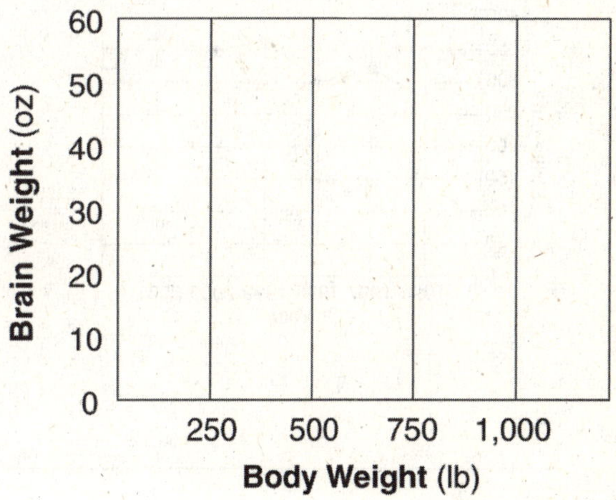

1. What kind of correlation do you see in the scatter plot?
   A positive correlation
   B negative correlation
   C no correlation

2. A dolphin's brain is 1% of its body weight. A human's brain is about 2.4% of its body weight. A cow's brain is about 0.1% of its body weight. Which animal has the biggest brain in relation to its size?
   A dolphin
   B human
   C cow

3. A white shark's brain is about 0.008% of its body weight. An alligator's brain is about 0.1% of its body weight. A sheep's brain is about 0.21% of its body weight. Which animal has the smallest brain in relation to its size?
   A white shark
   B alligator
   C sheep

4. Based on the information in the scatter plot, does an animal with a bigger body always have a bigger brain?
   A yes
   B no
   C not enough information

How do you count a herd of cows?

With a ____ ____ W C ____ L ____ ____ ____ R.

Name _______________ Date _______________ Class _______________

LESSON 7-10

Practice A

Misleading Graphs

1. Which graph could be misleading? Why?

Graph A

Graph B

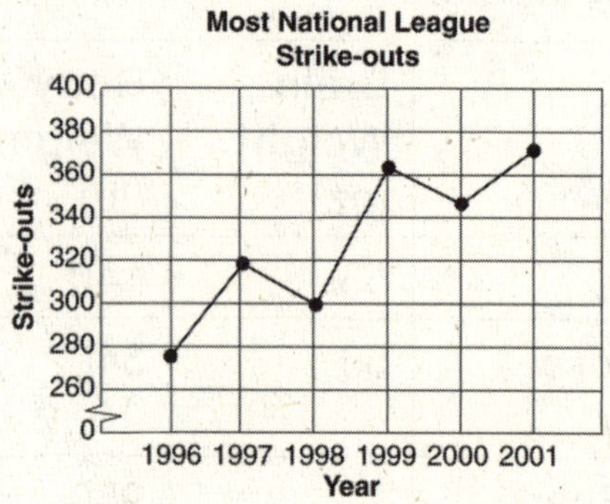

____

____

Tell why each graph could be misleading.

2.

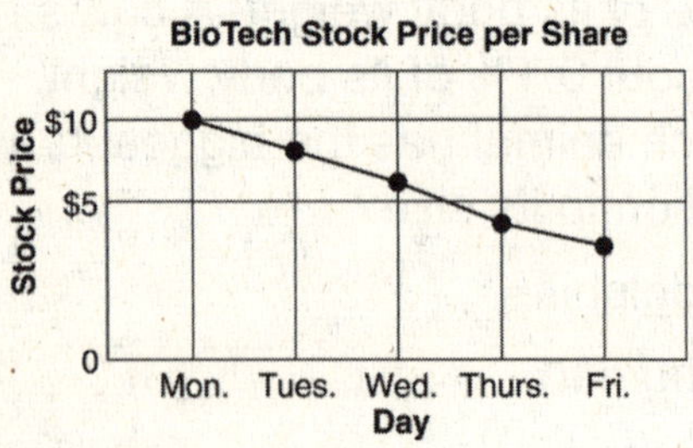

____

____

3.

____

____

Name ______________________ Date ______________ Class ______________

LESSON 7-10

Practice B

Misleading Graphs

1. Which graph could be misleading? Why?

Graph A

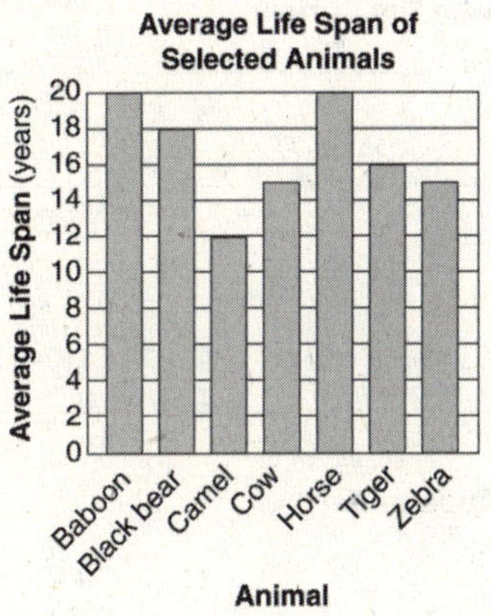

Graph B

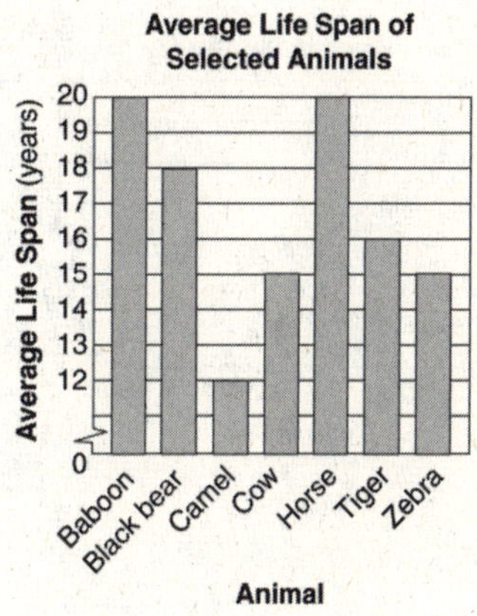

____

____

Explain why each graph could be misleading.

2.

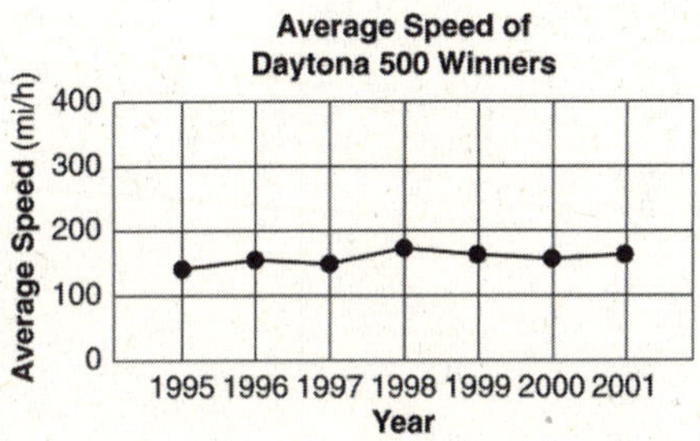

____

____

____

3.

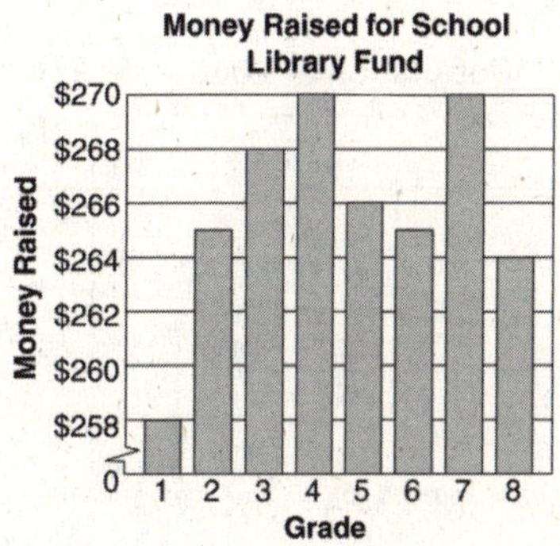

____

____

____

Name ______________________ Date ______________ Class ______________

LESSON 7-10

Practice C

Misleading Graphs

1. Which graph could be misleading? Why?

Graph A

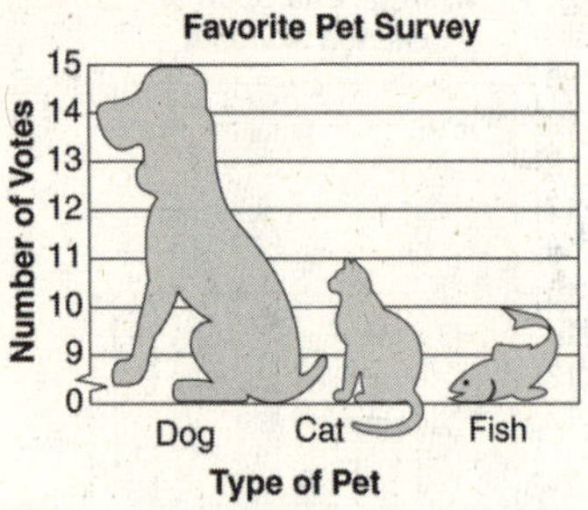

Graph B

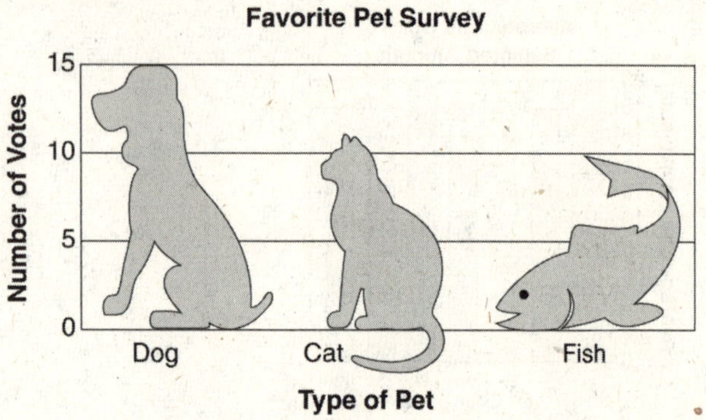

____

____

Explain why this graph could be misleading. Then redraw it so it is not misleading.

2.

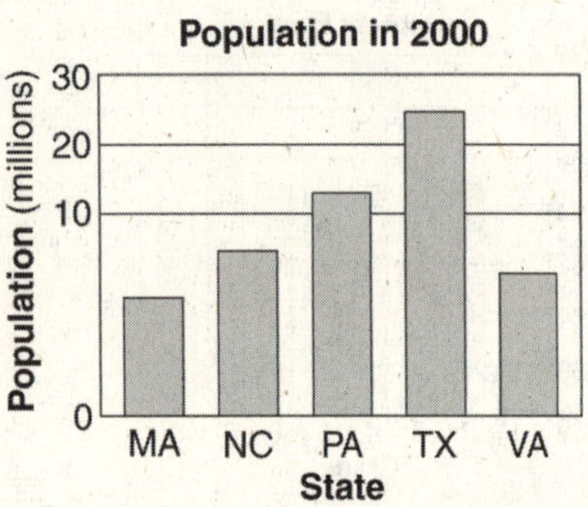

____

Name ______________________ Date ______________ Class ______________

LESSON 7-10

Review for Mastery

Misleading Graphs

Here are some ways a graph may be misleading.

- An axis is "broken" or numbers are left off an axis.
- The intervals on an axis are not the same length.
- Different sizes are used to represent bars in a bar graph.
- Pictorial graphs could distort the data.

Compare these graphs.

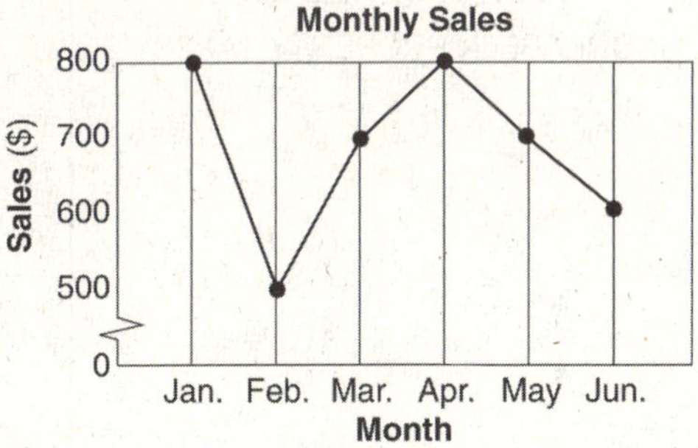

It looks like sales dropped by more than half from January to February. Look at the vertical axis. It is broken, so the scale does not start at 0. This graph is misleading.

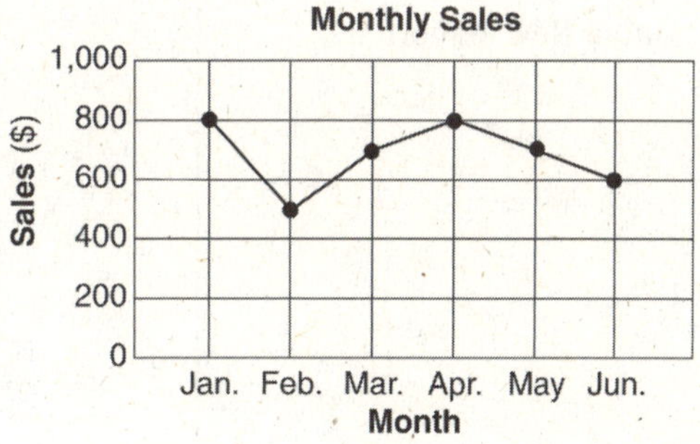

Look at the scale on the vertical axis. All the numbers from 0 to 1,000 are represented. This is a more realistic graph.

Explain why each graph is misleading.

1.

____
____
____
____

2.

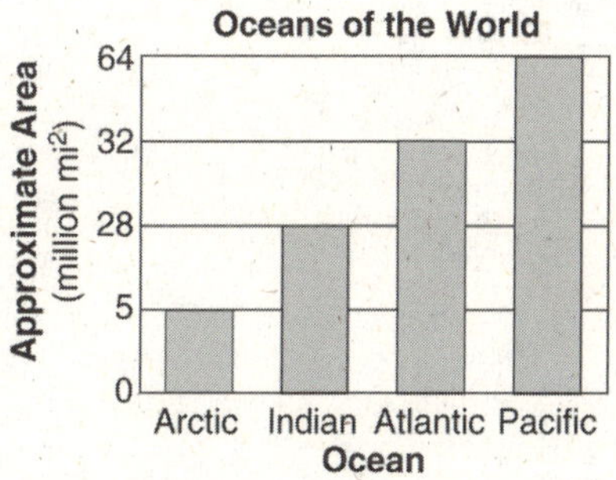

____
____

Name ______________________ Date ____________ Class ____________

LESSON 7-10

Challenge

Picture the Headlines

Read each fact. Draw a bar or line graph to support the two different headlines that report the fact.

1. Fact: In 2002, property taxes were $2.50 per $100 of the value of a house in Oakville. In 2003, they will increase to $2.60 for each $100 of value.

Property Taxes Skyrocket Under New Mayor!

NEW MAYOR OKAYS SMALL INCREASE IN PROPERTY TAXES

2. Fact: From 1996 to 1997, public schools in the United States averaged 7.8 students per computer. From 1997 to 1998, this average went down to 6.1. From 1998 to 1999, it was 5.7; and from 1999 to 2000, it was 5.4.

Schools See Big Drop in Number of Students per Computer Since 1996

Little Change Since 1996 in Number of Students per Computer in Our Public Schools!

Name ______________________ Date ______________ Class ______________

LESSON 7-10

Problem Solving

Misleading Graphs

Write the correct answer. Use the line graph for Exercises 1–4.

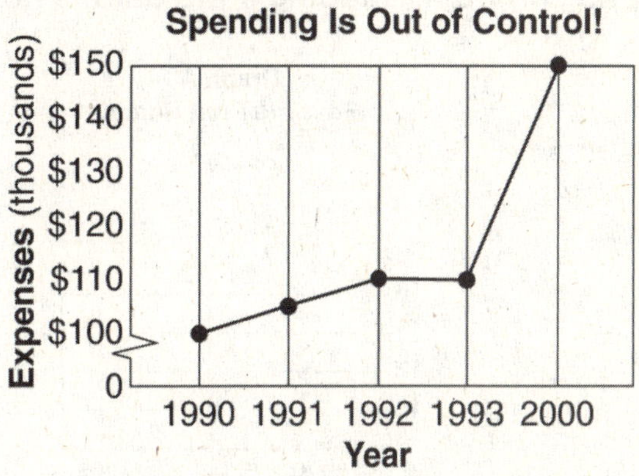

1. What would be a less misleading title for this graph?

 ____

 ____

2. How is the horizontal axis misleading?

 ____

 ____

3. How is the vertical axis misleading?

 ____

4. How much did spending really increase between 1993 and 2000?

 ____

Choose the letter for the best answer.

The bar graph is an advertisement used by a tour company to convince New Yorkers to vacation in Hawaii.

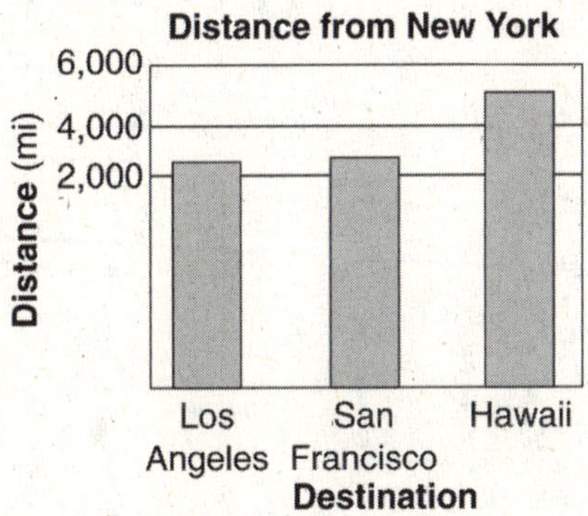

5. How far is New York from Los Angeles?
 - A 2,500 miles
 - B 4,000 miles
 - C 5,000 miles
 - D 5,200 miles

6. How far is New York from Hawaii?
 - F 2,500 miles
 - G 2,600 miles
 - H 5,000 miles
 - J 6,000 miles

7. What is the point of the ad?
 - A New York is closer to Hawaii than to Los Angeles.
 - B Hawaii is the same distance from New York as Los Angeles.
 - C Hawaii is only slightly farther from New York than is Los Angeles.
 - D San Francisco and Los Angeles are the same distance from New York.

8. Why is the graph misleading?
 - F The distances are incorrect.
 - G The bars are mislabeled.
 - H The bars are too tall.
 - J The intervals on the vertical axis are not equal.

Name ____________________ Date ____________ Class ____________

LESSON 7-10

Reading Strategies

Analyze Information

Graphs can be misleading if the data is not shown properly.

Graph A
Favorite Fruit Snack

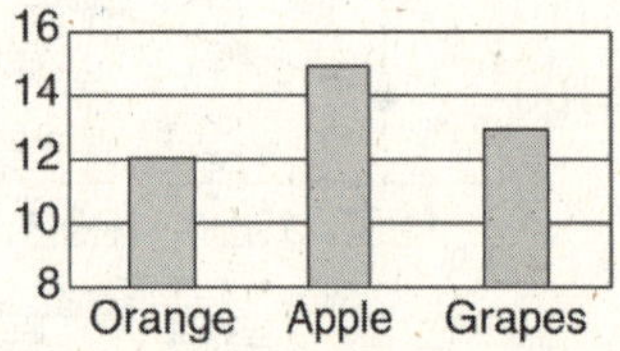

Graph B
Favorite Fruit Snack

Answer the following questions about Graphs A and B.

1. What number does the scale start on for Graph A? ____________________

2. What number does the scale start on for Graph B? ____________________

3. Which of the graphs is misleading? Why?

____

____

A graph is misleading if the intervals are too large or too small.

Graph C
Dinners Sold in One Month

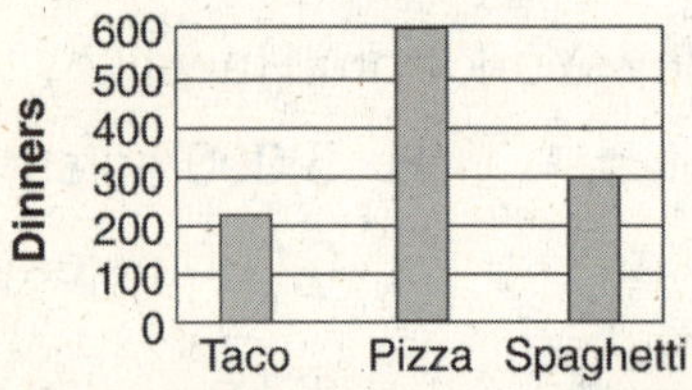

Graph D
Dinners Sold in One Month

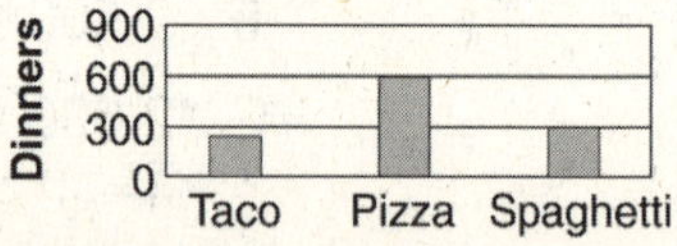

Answer the following questions about Graphs C and D.

4. What intervals are used in Graph C? ____________________

5. What intervals are used in Graph D? ____________________

6. Which graph is misleading? Why?

____

____

Name ______________________ Date ______________ Class ______________

LESSON 7-10

Puzzles, Twisters & Teasers

How's Your Vocabulary?

Complete the crossword puzzle by answering the clues.

Across

2. Each _____ or slice of a circle represents one part of the entire data set.
5. Each point on a scatter _____ represents a pair of data values.
7. bar and line _____
8. Frequency and cumulative frequency are two types of _____.
11. The _____ is the value or values that occur most often in a data set.
12. A _____ graph shows change over time for two sets of data.
13. The _____ is the sum of the data values divided by the number of data items.
14. The difference between the least and greatest values is called the _____.

Down

1. Often researchers can't survey every member of a large group, so they study a part of the group. This group is called a _____.
3. Positive, negative, or no _____ are three ways to describe data in a scatter plot.
4. the singular form of data
6. An extreme value in a set is called an _____.
9. The _____ is the middle value of an odd number of items arranged in order and it is the average of the two middle values for an even number of items.
10. A _____ table is one way to organize data into categories or groups.

Answers

LESSON 7-1

Practice A

1.

Normal Monthly Temperatures in Tampa		
Temperature (°F)	Frequency	Cumulative Frequency
60–69	5	5
70–79	3	8
80–89	4	12

2. 8

3.

Stems	Leaves
6	1 3 3 7 9
7	2 6 8
8	2 2 3 3

Key: 6 | 1 means 61

4. 5

5. 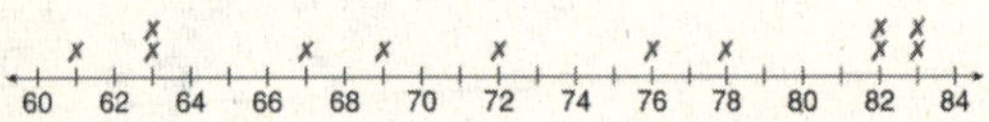

Practice B

1.

Heights of Students		
Height (in.)	Frequency	Cumulative Frequency
45–49	5	5
50–54	5	10
55–59	9	19
60–64	5	24

2. 19

3. **Heights of Students**

Stem	Leaves
4	5 7 8 8 9
5	1 2 2 3 4 5 5 6 6 7 7 8 8 9
6	0 0 0 1 2

Key: 5 | 2 means 52

4. 5

5. 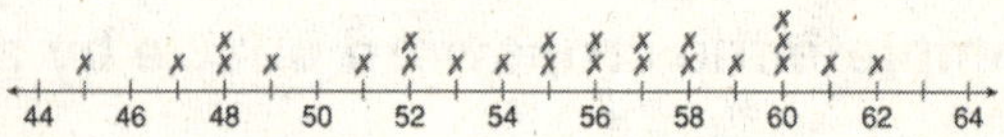

6. 60 in.

Practice C

1.

Ages of Twentieth Century Presidents		
Ages	Frequency	Cumulative Frequency
40–44	2	2
45–49	1	3
50–54	5	8
55–59	4	12
60–64	4	16
65–69	1	17

2. **Ages of Twentieth Century Presidents**

Stems	Leaves
4	2 3 6
5	1 1 1 2 4 5 5 6 6
6	0 1 2 4 9

Key: 4 | 6 means 46

3. 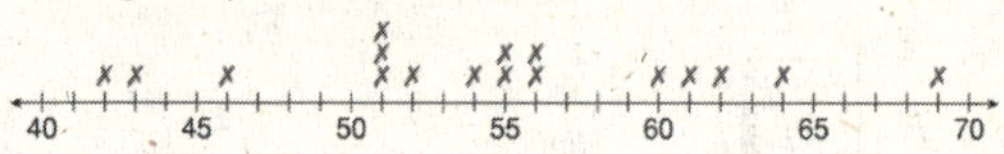

4. 9

5. Possible answer: b; Cars on a highway could be traveling between 51 and 72 miles per hour. Classes probably would not have 51–72 students, and part-time employees would not be working that many hours.

Review for Mastery

1.

Highest Temperatures in M and N States		
Temperature (°F)	Frequency	Cumulative Frequency
105–109	5	5
110–114	4	9
115–119	4	13
120–124	2	15
125–129	1	16

2.

Tallest Buildings in Houston

Stems	Leaves
3	6 6
4	2 4 5 6 6 7 7 9 9
5	0 0 2 3 5 6
6	4
7	1 5

Key: 3 | 6 means 36

Challenge

1. 76 °F; 78 °F
2. Bloomington
3. Bloomington
4. Possible answer: Bloomington; It has more low temperatures.
5. 40 °F
6. 4 months
7. 27 °F
8. Richmond's highest average temperature was 2° higher than Bloomington's.

Problem Solving

1. 4
2. 10, 25, 30, 52
3. stem 2; leaves 0, 1, 2, 4, 5, 5
4. Stems 4 and 5 each have 3 leaves.
5. B
6. F
7. D
8. H

Reading Strategies

1. in intervals of 10
2. 3
3. 6
4. left
5. 4
6. 36

Puzzles, Twisters & Teasers

Minutes	Frequency	Cumulative Frequency	
10–20	4	4	A
21–30	2	6	E
31–40	3	9	L
41–50	2	11	S
51–60	3	14	V

L E A V E S

Stems	Leaves	
1	0 0 5 8	K
2	5 8	N
3	2 5 5	R
4	5	T
5	0 5 7 9	U

T R U N K

LESSON 7-2

Practice A

1. 6; 5; 5; 6
2. 56; 56; no mode; 8
3. 20; 18; 18; 9
4. 35; 35; no mode; 50
5. 7; 5.5; 5 and 6
6. Possible answer: The median best describes the data set because it is closer than the mean to the number of kilometers that Clara ran the majority of the days. There are two modes.
7. 61
8. It increases the mean by 7 and the median by 1. It does not affect the mode, because there is no mode.

9. Possible answer: The median best describes the data set with the outlier.
10. Possible answer: The mean and median best describes the data set without the outlier.

Practice B

1. 38; 37; no mode; 30
2. 60; 56.5; 56; 28
3. 25; 19; 19; 70
4. 9.6; 8.5; 8; 14
5. The median best describes data set because it is closest to the numbers of hours reported by most of the students.
6. Outlier, 5; It decreased the mean by $1.\overline{3}$ and the median by 0.5. It did not affect the mode. With the outlier, the data is best described by the median. Without the outlier, the data is best described by the mean.
7. Outlier, 92; It increased the mean by 4.25 and the median by 2. It did not affect the mode. With the outlier, the data is best described by the median. Without the outlier, the data is best described by the mean.

Practice C

1. 17; 16.5; 17; 10
2. 53; 56; 56; 24
3. Possible answer: The median best describes data set because it is closest to the majority of the point totals.
4. Outlier, 10; It decreased the mean by 2.5 and the median 2. It did not affect the mode. With the outlier, the data is best described by the median. Without the outlier, the data is best described by the mean or the median.
5. Outlier, 103; It increased the mean by 4.75 and the median by 0.5. It did not affect the mode. With the outlier, the data is best described by the median or the mode. Without the outlier, the data is best described by the mean.
6. mode; Possible answer: No; the mode was higher than all of the other scores, and occurred only one more time than the score that occurred the next greatest number of times.

Review for Mastery

1. 5; 5.6; 6; 6
2. 8; 14; 15; 17
3. 16; 25.5; 24.5; 23
4. 10; 11; 11; no mode
5. 7; 42; 42; 42
6. 17; 61; 62; 62 & 68
7. 11 °F; 42 °F; 42 °F; no mode
8. It increases the mean by 3.8.
9. It increases the median by 1.
10. There is no effect.
11. Yes; 25
12. It increases the mean by 3 and the median by 0.5.
13. The median best describes the data set because it is least affected by the outlier.

Challenge

1. Possible answer: 8, 10, 11, 11, 12, 14
2. 10, 12, 12, 16, 17, 18, 20
3. 8, 9, 9, 10, 14, 14, 15, 17
4. 14
5. 8
6. 33
7. 27
8. 91
9. He can't do it. He would need to score about 108 on each quiz.

Problem Solving

1. 42
2. 69.4; 63; there is no mode
3. 100
4. It increases the mean by 7.65 and the median by 0.5.
5. Possible answer: The median; the outlier affects the median less than it does the mean.

6. A
7. G
8. C
9. G

Reading Strategies

1. mode
2. mean
3. range
4. median

Puzzles, Twisters & Teasers

T: 8
L: 4
I: 3.5
U: 11
R: 5.25
O: 5
E: 2

O U T L I E R

LESSON 7-3

Practice A

1. Nile River
2. about 1,100 mi
3. about 800 mi
4.

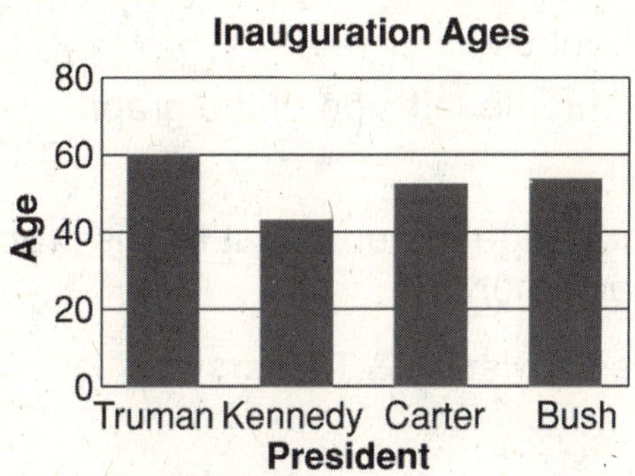

5.

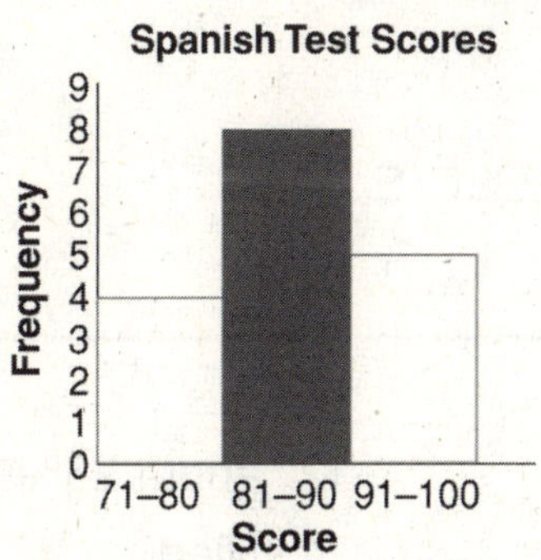

Practice B

1. Alaska
2. about 4,000 ft
3. about 9,000 ft
4.

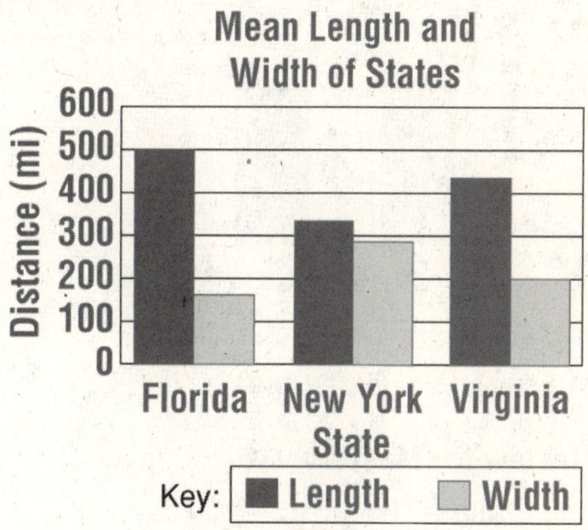

5.

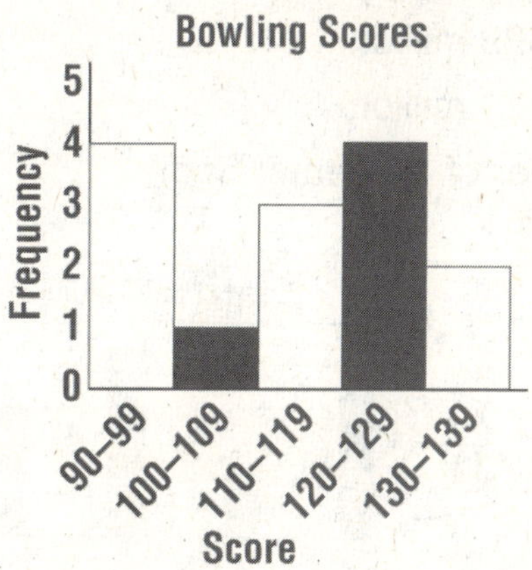

Practice C

1. bicycling
2. about 225,000 children
3. about 20,000 children
4.

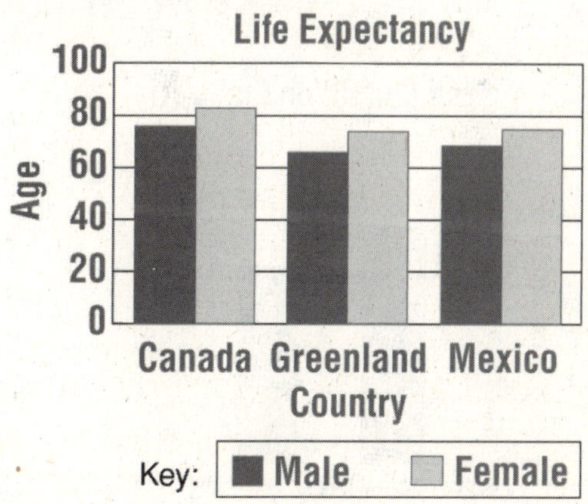

5.

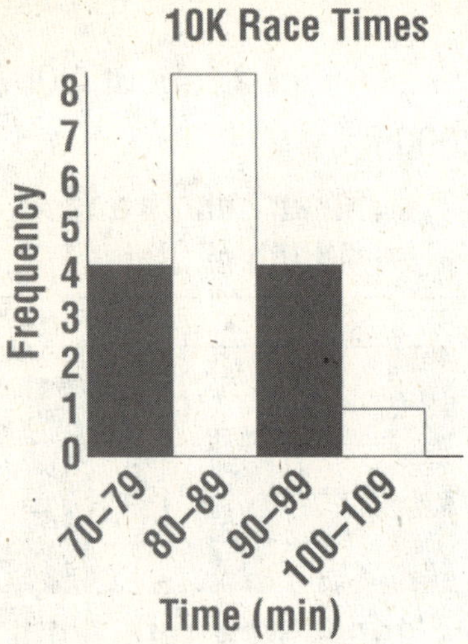

Review for Mastery

1. Dallas
2. about 29 million
3. about 11 million

4a. Number of Students; Sport

4a-c.

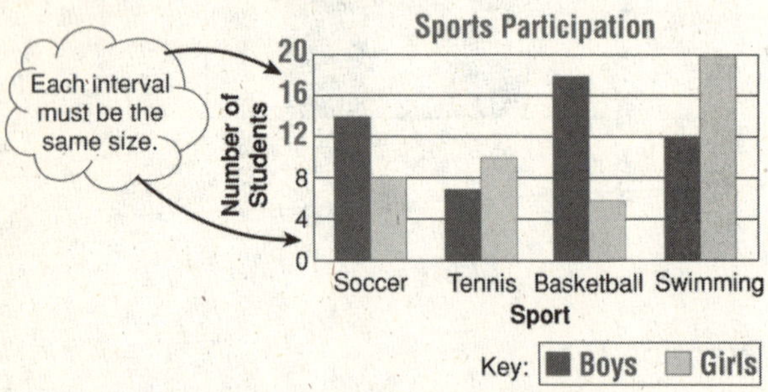

5. 5 patients
6. 90–99
7. 3 more patients

8a. Frequency; Fish Length (in.)

8a-d.

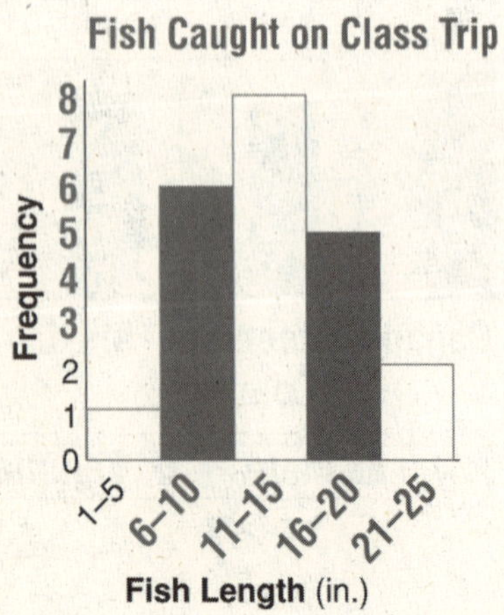

Challenge

1. 5 years
2. horizontal
3. females
4. about 3.3 million
5. about 3.6 million
6. 40–44; 40–44
7. about 4.3 million
8. about 0.6 million
9. about 0.5 million
10. age 75 and above

Problem Solving

1. 5
2. 2003
3. between 2001 and 2002
4. 2001
5. A
6. J
7. C
8. F

Reading Strategies

1. boys' favorite sports
2. tens
3. about 25
4. about 54
5. along the left side of the graph
6. 10
7. the range of hours that teams practice each month

Puzzles, Twisters & Teasers

C: 1

E: 2

L: 3

I L L – E A G L E A C T

LESSON 7-4

Practice A

1. apples
2. about 25%
3. 5 teens
4. filled
5. about 50%
6. 30 people
7. a bar graph; One axis could show the number of bears, and the other axis could show the parks.

8. a circle graph; The circle represents all summer visitors.

Practice B

1. piano
2. about 25%
3. 16 teens
4. beaches
5. about 25%
6. 15 people
7. a bar graph; One axis could show the number of tornadoes and the other could show the states.
8. a circle graph; The sector for the Macintosh apples could be compared to the entire circle.

Practice C

1. baseball
2. about 10%
3. 24 teens
4. cereal
5. about 20%
6. 50 people
7. a circle graph; The sector for entertainment could be compared to the entire circle.
8. a bar graph; One axis could show the number of calories, and the other could show the birds.

Review for Mastery

1. snowboarding
2. about 25%
3. 25 teens
4. bar graph; One axis could show the number of hurricanes, and the other could show the states.
5. circle graph; The sector for those who play soccer could be compared to the entire circle.

Challenge

1. **Oceans of the World**

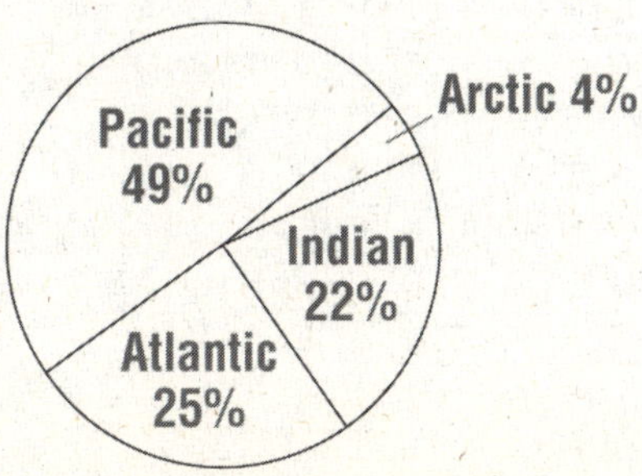

2. about 64 million mi^2
3. about 33 million mi^2
4. about 29 million mi^2
5. about 5 million mi^2
6. about 96 million mi^2
7. about 3 million mi^2

Problem Solving

1. $\frac{1}{2}$ of the circle
2. $\frac{1}{4}$ of the circle
3. circle graph
4. bar graph
5. C
6. H
7. B
8. J

Reading Strategies

1. Molly's 60-minute exercise program
2. aerobics
3. Half of the circle is labeled "aerobics."
4. toning exercises and stretching exercises
5. They are equal sections of the circle.
6. 60 minutes
7. 30 minutes
8. 15 minutes

Puzzles, Twisters & Teasers

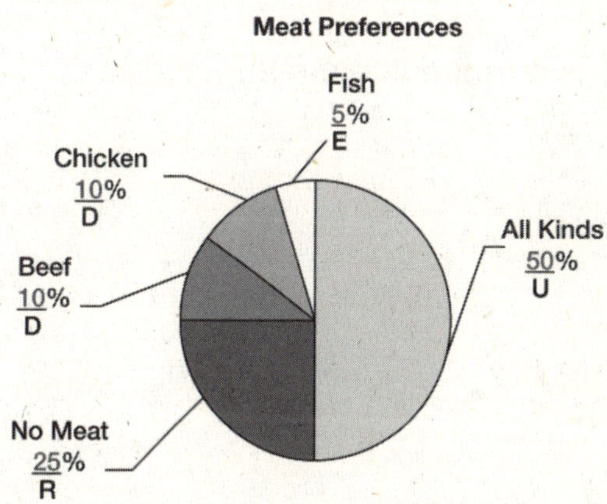

U D D E R

LESSON 7-5

Practice A

1.

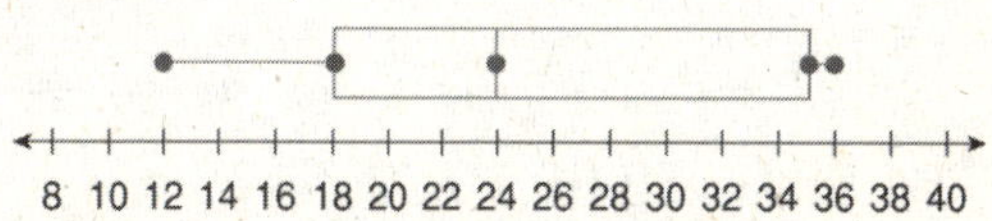

2. Amy
3. Ed
4. Ed
5. Amy; Possible answer: The range and interquartile range are smaller for Amy than for Ed, so Amy's test scores are more predictable.
6. Port Eagle
7. Port Eagle
8. Surfside; Possible answer: The interquartile range is smaller for Surfside than for Port Eagle, so Surfside's room prices are more predictable.

Practice B

1. 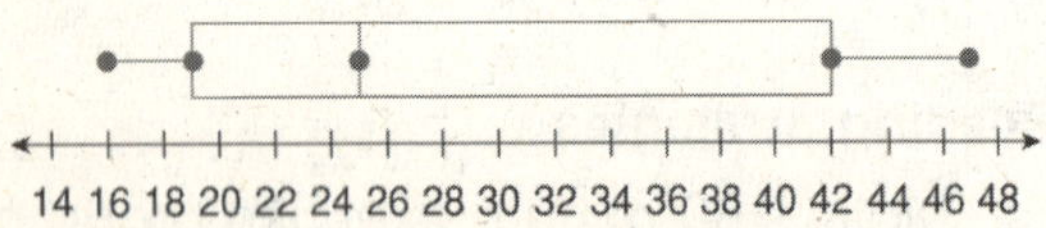

2. New York
3. Arizona
4. New York
5. Greystone Inn
6. Greystone Inn
7. Sam's Place

Practice C

1. 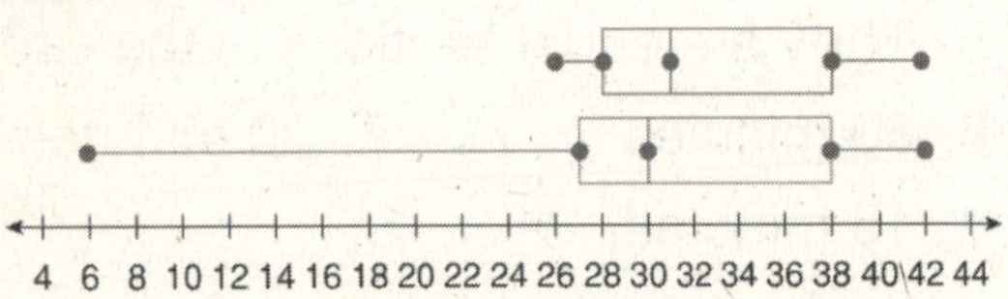

2. It increases the interquartile range by 1.
3. The range is more affected.
4. 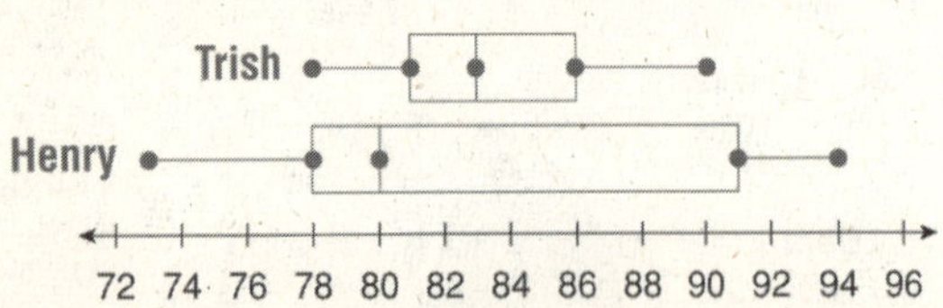

5. Henry
6. Trish
7. Trish

Review for Mastery

1. 20 24 25 27 31 35 38
2. 20, 38, and 27
3. 24, 35
4. 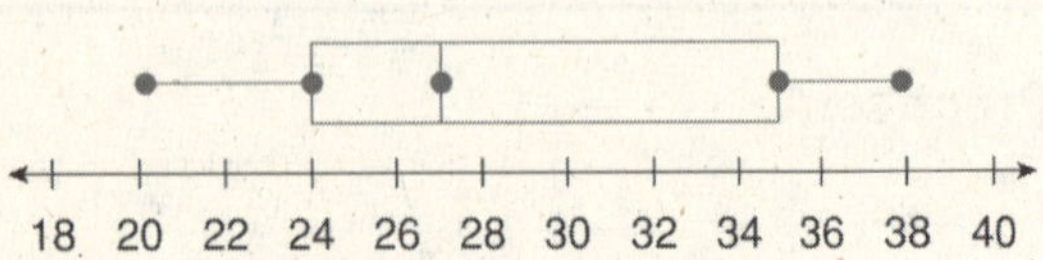

5. 61 63 65 68 69 70 72 74 78
6. 61, 78
7. 69, 64, and 73
8. 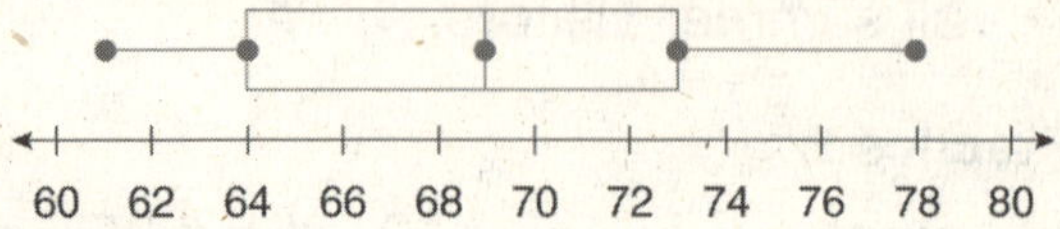

Challenge

1. 10; 24
2. 18
3. The median is a number in the data set. When there are an odd number of numbers, the median is the middle number.
4. 12, 23
5. Yes. Since the lower and upper halves of the data each have an odd number of numbers, the median of each is the middle number in each.
6. 16, 22
7. 6, 9, 10, 12, 14, 15, 20

Problem Solving

1. Class B; 8
2. Class B
3. Class A
4. 25%
5. B
6. H
7. D
8. J

Reading Strategies

1. the middle half of the data
2. The whiskers extend from both ends of the box to the least and greatest values.
3. It divides the set of data in half.
4. least value; lower quartile
5. upper quartile; greatest value
6. median; upper quartile

Puzzles, Twisters & Teasers

E	B	O	X	Z	D	W	P	Y	J	I	O	R	C
X	F	D	Q	O	W	E	R	F	C	O	D	A	T
P	R	R	U	E	H	H	T	I	P	J	A	D	A
L	A	D	A	W	T	H	I	R	D	U	T	B	M
O	C	A	R	C	S	B	I	S	R	P	A	I	E
T	T	O	T	A	T	A	L	T	K	P	G	H	D
U	I	K	I	O	N	M	T	N	F	E	A	Z	I
P	O	M	L	D	E	G	R	D	G	R	R	W	A
L	N	B	E	X	T	R	E	M	E	E	S	V	N

M T (empty)

LESSON 7-6

Practice A

1.

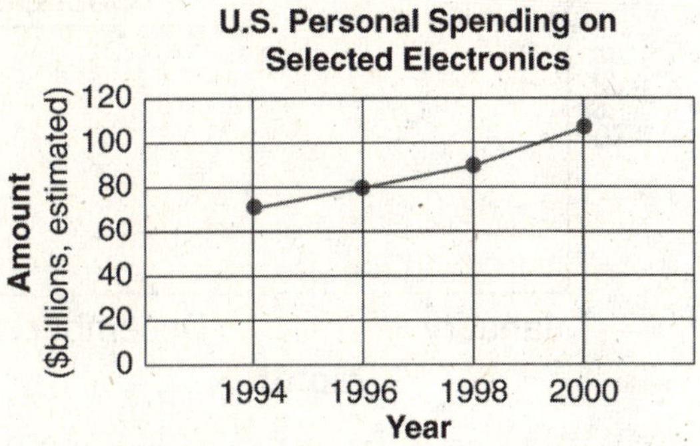

2. from 1998 to 2000
3. Possible answer: about $100 billion
4.

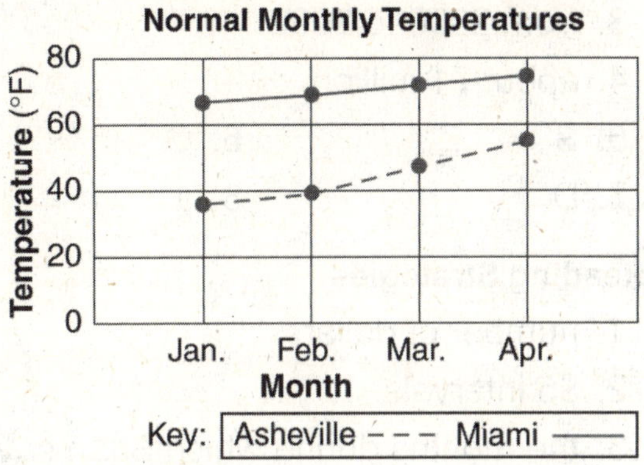

5. during January
6. during April
7. Miami
8. Asheville

Practice B

1.

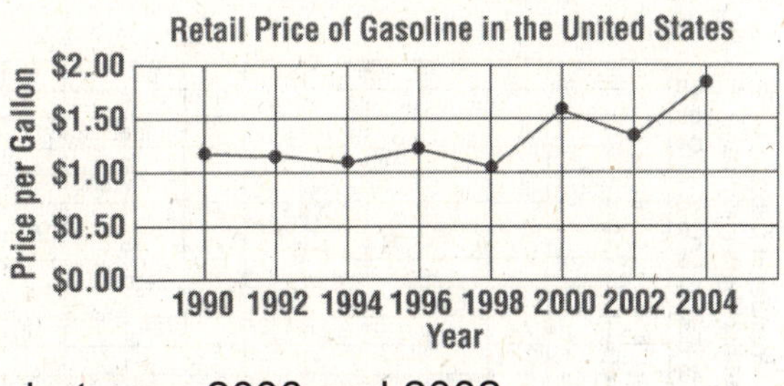

2. between 2000 and 2002
3. Possible answer: about $1.17
4.

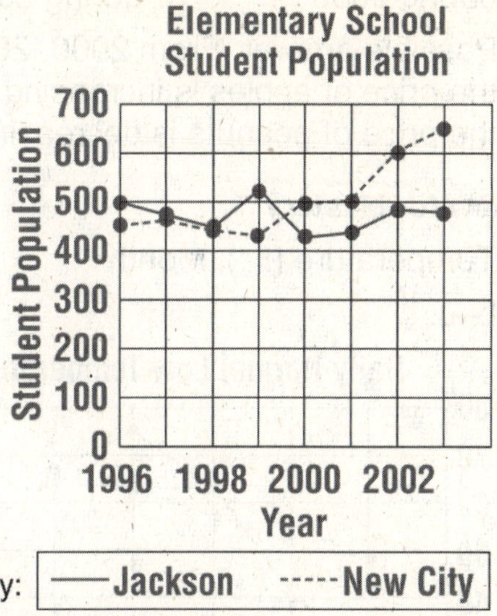

5. from 2001 to 2002
6. 2000; Possible answer: The student population jumped in New City, but dropped in Jackson.

Practice C

1.

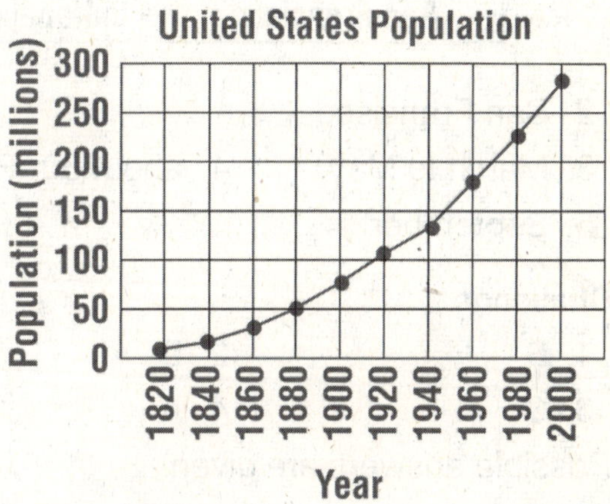

2. from 1980 to 2000
3. about 200 million
4.

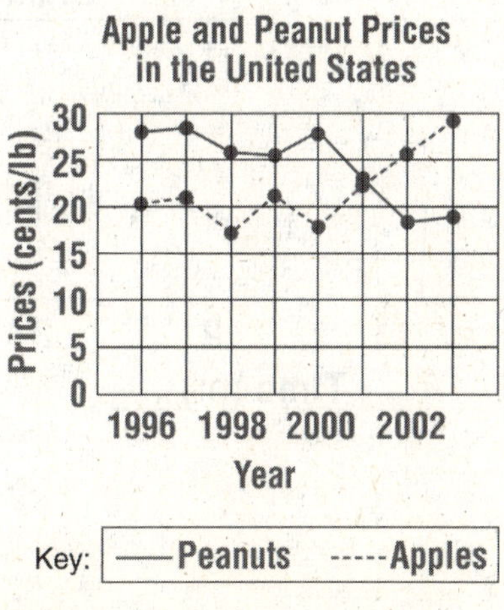

5. during 2003
6. during 2001
7. Possible answer: From 2000–2003, the price of apples is increasing and the price of peanuts is decreasing.

Review for Mastery

1. Temperature (F°); Month

   b–d.

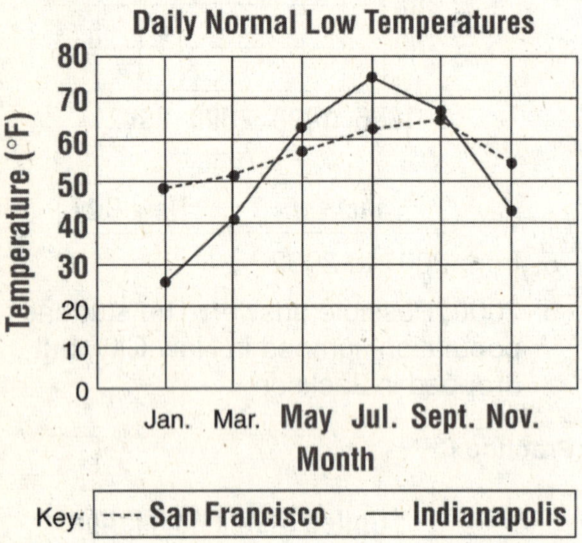

2. San Francisco
3. March to May
4. about 60 °F
5. September

Challenge

1. B
2. D
3. C
4. A

Possible answers are given.

5.

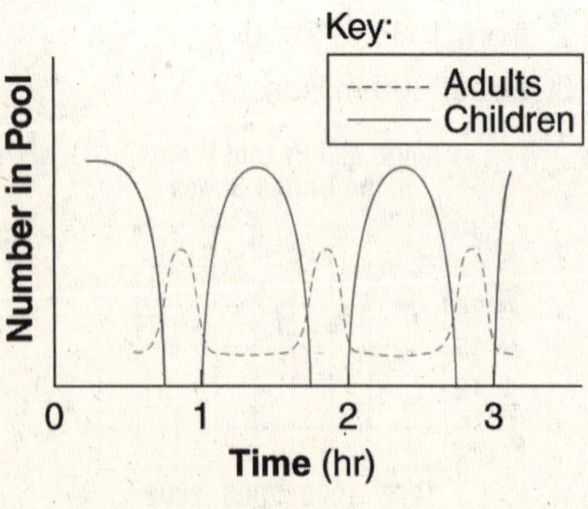

6.

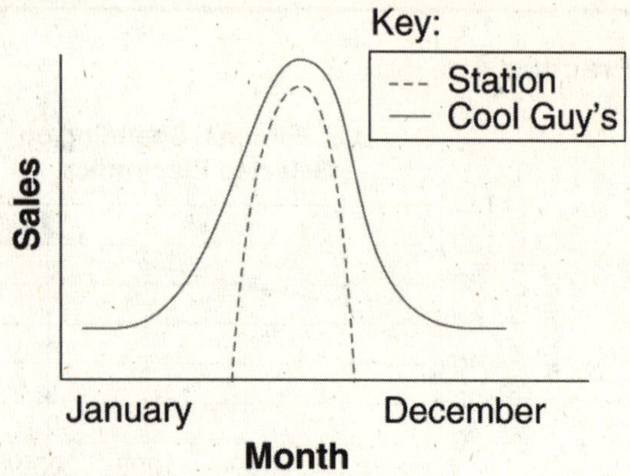

Problem Solving

1. about 67 million
2. about 9 million
3. from 2000 to 2002
4. about 71 million
5. B
6. G
7. D
8. J

Reading Strategies

1. number of dollars
2. $5 intervals
3. the months during which Brian saved money
4. $15
5. $15
6. Possible answer: He spent some of his savings.

Puzzles, Twisters & Teasers

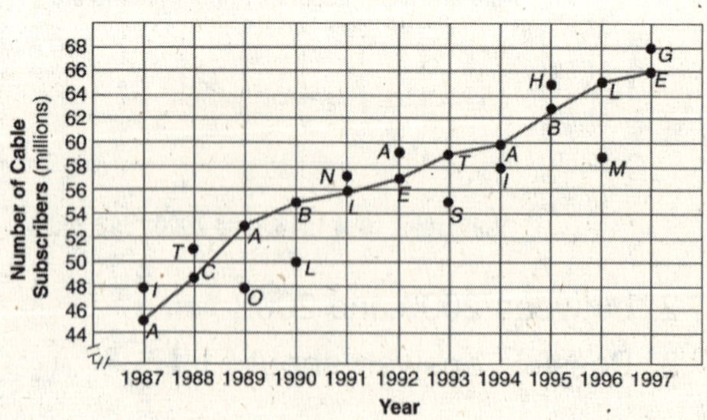

Year	Number of Cable Subscribers (millions)	Letter to Solve Riddle
1987	45	A
1988	49	C
1989	53	A
1990	55	B
1991	56	L
1992	57	E
1993	59	T
1994	60	A
1995	63	B
1996	65	L
1997	66	E

LESSON 7-7

Practice A

1. D
2. H
3. A
4. G
5. Possible answer: No. There are only four values, and how often they occur and how they are distributed is not important.
6. Yes. The bar graph shows how many miles each student ran, and makes it easy to compare the amounts.

Practice B

1. Possible answer: bar graph
2. Possible answers: stem-and-leaf graph or line plot
3. Possible answer: Venn diagram
4. Possible answer: circle graph
5. A circle graph displays data as parts of a whole. Since data in this table do not represent parts of whole, a circle graph is not appropriate.
6. The table shows data that changes over time. Since a line graph shows how data changes over time, this graph is appropriate.

Practice C

1. possible answers: stem-and-leaf plot or line plot
2. possible answers: line graph or bar graph
3. A circle graph shows how data is divided into parts. Since this circle graph shows the part of the total number of votes that each sport received, it is appropriate.
4. A line plot shows how often data values occur and how they are distributed. Since the frequency and distribution of the data in the table is not important, this graph is not appropriate.
5. Check students' graphs. Possible answers: a line plot or a stem-and-leaf plot.

Review for Mastery

Possible answers are given.

1. circle graph

   The vote is split into parts. A circle graph shows parts of a whole.
2. stem-and-leaf plot

   The display should show each point total. A stem-and-leaf plot shows each piece of data.
3. line graph

   The data involves amounts changing over time. A line graph shows amounts changing over time.
4. bar graph

   The display should compare the four pieces of data. A bar graph compares data.

Challenge

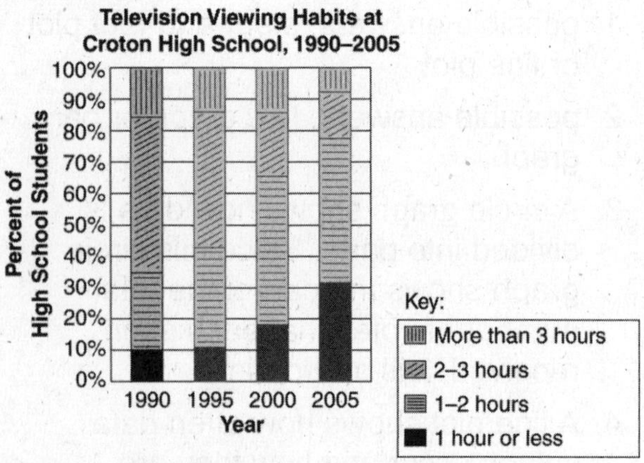

Problem Solving

1. Possible answer: Line plot; a line plot shows each piece of data and how the data are distributed.
2. Possible answer: Circle graph; the parts of the budget represent parts of a whole, and a circle graph shows parts of a whole.
3. Possible answer: Since the data ranges from 40–108, a line plot would be very long.
4. Possible answer: No; a line graph displays changes over time, and this data does not involve time.
5. C   6. G

Reading Strategies

1. Do you need to show parts of a whole?
2. circle graph
3. Do you need to see how often each piece of data occurs?
4. line plot or stem-and-leaf plot

Puzzles, Twisters & Teasers

1. O
2. Y
3. G
4. E
5. D
6. S

GOOD EYES

LESSON 7-8

Practice A

1. b   2. a
3. The sample is biased. Possible explanation: It is likely that people who use the library will not have the same opinion as people who do not use the library.
4. The sample is not biased. It is a random sample.
5. The sample is biased. Possible explanation: It is likely that students who are not on sports teams will not have the same opinion as students who are on sports teams.
6. Yes; $\frac{2}{500} = \frac{48}{12{,}000}$, and $48 < 50$

Practice B

1. Suki's method will better represent the entire student population because she uses a random sample. Dinah's method produces results that are not as representative of the entire population because she uses a convenience sample.
2. The sample is not biased. It is a random sample.
3. The sample is biased. Possible explanation: It is likely that not all high school students will choose the same kind of music as the kind chosen by those who attend a jazz concert.
4. Yes; $\frac{7}{50} = \frac{56}{400}$, and $56 > 50$
5. 1,125 coins will have misprints.

Practice C

1. Wanda's method will better represent the entire employee population because she uses a random sample. Bernard's method will produce results that are not as representative of the entire population because his sample is not random.

2. The sample is biased. Possible explanation: It is likely that not all tenants will give the same answer as those tenants who use e-mail.
3. The sample is not biased. It is a random sample.
4. Possible answer: You would survey the entire population because the population is small.
5. Possible answer: You would use a sample because the entire population is large.

Review for Mastery

1. The sample is biased. The passengers on one on-time flight are likely to feel differently about their flight than passengers on some other flights.
2. The sample is not biased. It is a random sample.
3. The sample is not biased. It is a random sample.
4. The sample is biased. The people who go to movies are not likely to spend amounts of money on entertainment that are similar to the amounts spent by other people in the state.

Challenge

1. systematic	2. cluster
3. stratified	4. stratified
5. cluster	6. stratified
7. systematic	8. cluster
9. systematic	10. stratified

Problem Solving

1. No; Not all children are likely to have the same exercise habits as children who go to a sporting goods store.
2. Possible answer: She should survey the entire population, since it is relatively small.

3. 300 fish	4. 540 students
5. D	6. F

Reading Strategies

1. Population is everyone and a sample is part of the population.
2. An unbiased sample represents the population and a biased sample does not.
3. biased sample
4. unbiased sample
5. biased sample

Puzzles, Twisters & Teasers

1. O
2. N
3. D
4. K
5. M

D O N O T S M O K E

LESSON 7-9

Practice A

1.

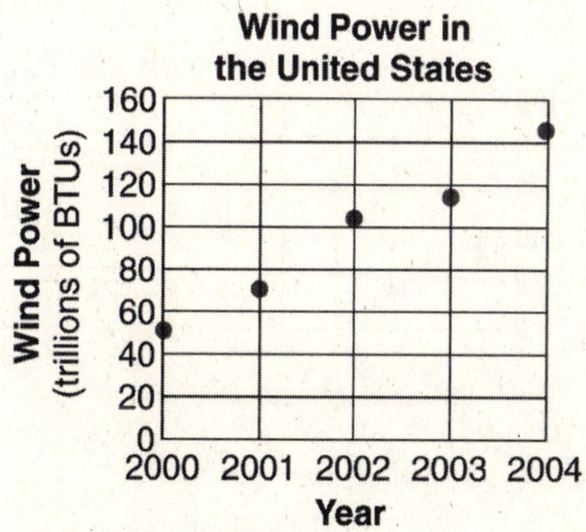

2. The amount of wind power increased over time.
3. positive correlation
4. negative correlation
5. no correlation
6. negative correlation
7. positive correlation
8. negative correlation
9. no correlation

Practice B

1.

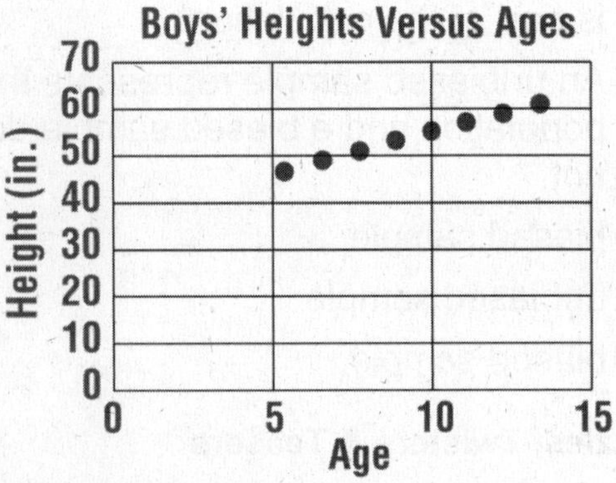

2. As boys get older, they grow taller.
3. positive correlation
4. negative correlation
5. positive correlation
6. no correlation
7. positive correlation
8. negative correlation
9. no correlation

Practice C

1.

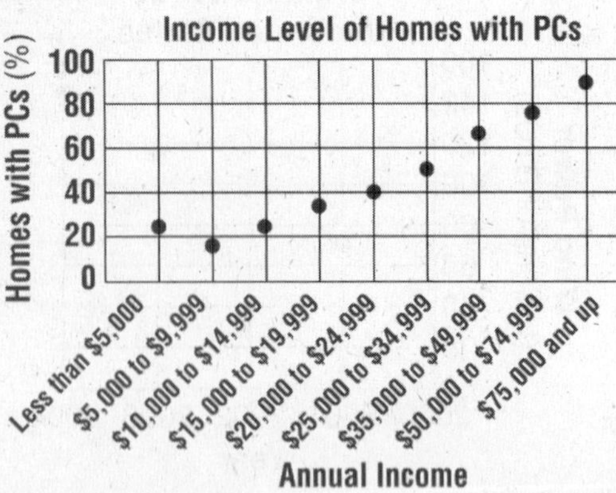

2. The higher the household income, the greater the percentage of homes with a PC.
3. positive
4. negative correlation
5. positive correlation
6. no correlation
7. positive correlation

Review for Mastery

1. positive correlation
2. no correlation
3. no correlation
4. negative correlation
5. positive correlation
6. negative correlation
7. no correlation
8. negative correlation

Challenge

Answers will vary.

1. Possible answer: negative
2. Possible answer: The more times I practiced writing my name backwards, the less time it took.
3.

Making tacos in a fast food restaurant	✓
Filling boxes on an assembly line	✓
Writing multiplication facts from 1 to 12	✓

4. Possible answer: The more you practice, the better and faster you will do the job.

Problem Solving

1. 16 years, B.A.
2. about $85,000
3. less than 12 years
4. positive correlation
5. C
6. G
7. A
8. H

Reading Strategies

1. Plot A; false
2. Plot B; true
3. Plot A; true
4. Plot C; false

Puzzles, Twisters & Teasers

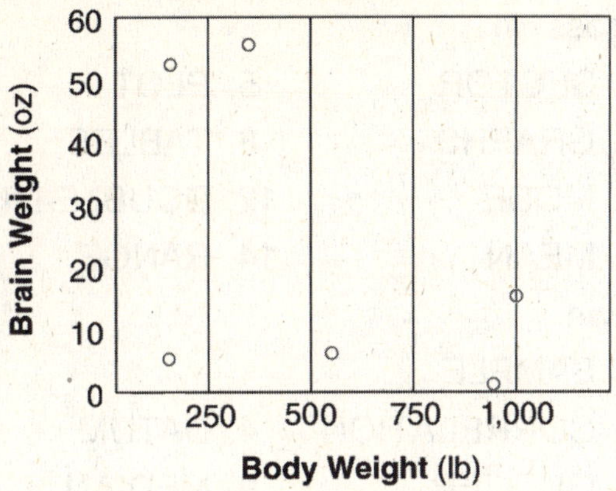

1. C
2. B
3. A
4. B

COWCULATOR

LESSON 7-10

Practice A

1. Graph B; Possible answer: The vertical axis is broken; it looks as if the number of strikeouts in 2001 is twice as great as in 1996.
2. The intervals on the vertical axis are unequal.
3. The broken axis makes Discount Mike's look much less expensive.

Practice B

1. Graph B; Possible answer: The vertical axis is broken; it looks as if the camel lives about $\frac{1}{4}$ as long as a horse, but that is not true.
2. The intervals are too great to show the difference in speeds.
3. The broken axis makes it look as if there is a large difference in the amount raised.

Practice C

1. Graph A: Possible answer: The drawing of the dog is so large that it looks as if the dog got twice as many votes as any other animal.
2.

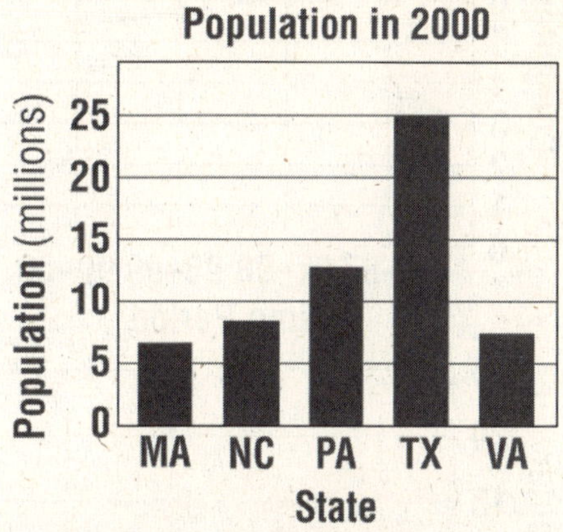

Unequal intervals make the populations look similar.

Review for Mastery

1. The skate showing sales at Bike and Blade is large, so it looks like the sales were more than twice as much.
2. The intervals on the vertical axis are not the same length.

Challenge

Possible answers are given.

1.

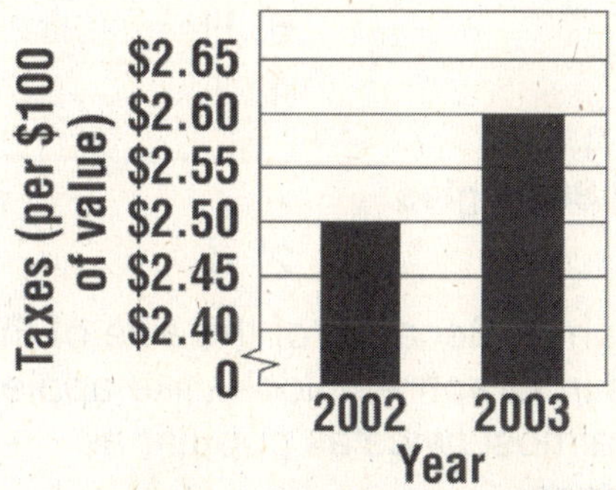

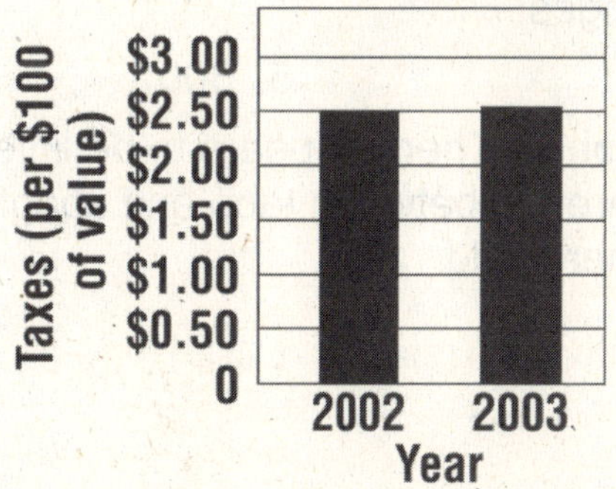

2.

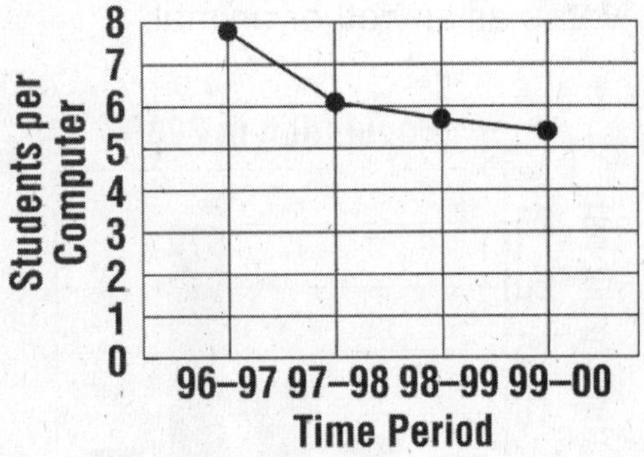

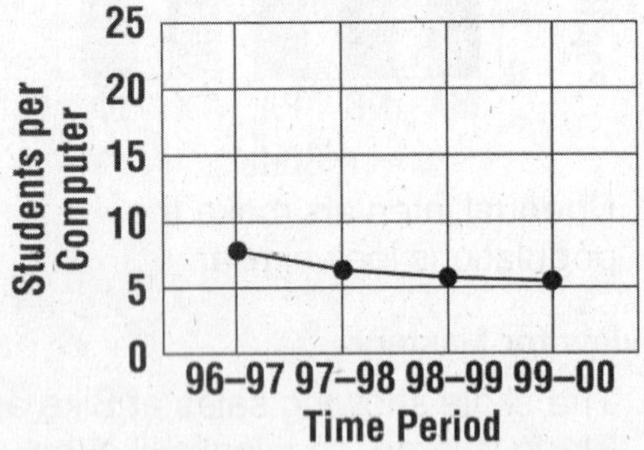

Problem Solving

1. Possible answer: Spending, 1990–2000
2. The intervals between years are not equal.
3. broken scale
4. about $40,000
5. A
6. H
7. C
8. J

Reading Strategies

1. 8
2. 0
3. Graph A; Because of the size of the bars in Graph A, it looks like apples are almost twice as popular as oranges.
4. 100
5. 300
6. Graph D; There appears to be little difference between taco and spaghetti dinners sold.

Puzzles, Twisters & Teasers

Across

2. SECTOR
5. PLOT
7. GRAPHS
8. TABLES
11. MODE
12. DOUBLE-LINE
13. MEAN
14. RANGE

Down

1. SAMPLE
3. CORRELATION
4. DATUM
6. OUTLIER
9. MEDIAN
10. FREQUENCY